Experimenter's Guide to Solid State Electronics Projects

Experimenter's Guide to Solid State Electronics Projects

Alfred W. Barber

Photographs by: Kyle H. Powell

Parker Publishing Company, Inc.

PARKER PUBLISHING COMPANY, INC.

West Nyack, N.Y.

Library of Congress Cataloging in Publication Data

Barber, Alfred W
Experimenter's guide to solid state electronics projects.

1. Solid state electronics--Handbooks, manuals, etc.
2. Solid state electronics--Experiments. I. Title.
TK7871.85.B374 621.3815 79-24002
ISBN 0-13-295451-6

Printed in the United States of America

How This Book Will Help the Experimenter

Experimenting with solid state electronics is a never ending program of exciting challenges. The excitement can start with a simple kit and continue with more complex and far more rewarding projects. It can be the source of much fun, provide an absorbing hobby, enable you to create useful and enjoyable products or enliven a professional career.

Experimenting will also help you confirm or disprove something doubtful or uncertain, discover some new effect or principal and test or find the best way to accomplish a desired result. While some experiments may follow prescribed steps to accomplish known ends, the greatest excitement comes from a discovery of something new or the perfection of something old to a degree not known before. For example, experiments will be described in which the efficiency of solar cells is optimized.

Solid state electronics has a very special appeal for the experimenter. The components are small, inexpensive and readily utilized in experiments. The results are often dramatic and can provide both useful devices and just fun things.

Solid state electronic devices include diodes, rectifiers, transistors, light-emitting diodes, photoelectric cells, fibre optics, liquid crystals, silicon controlled rectifiers and both linear and digital integrated circuits. These devices are used in electronic watches, automatic exposure cameras, radios, television receivers, pocket calculators, microprocessors, computers, light dimmers, automobile ignition systems, aviation equipment, military equipment and space exploration controls. They are also responsible for smaller devices like "pocket" radios, "pocket" calculators and video games; for expanding man's space-age capabilities, like going to the moon; for having a computer in your home; for communications capabilities such as satellites and fiber optics; for power from sunlight and seemingly impossible devices operated by remote control, like the probes on Mars and Venus.

The experiments in this book are both interesting and informative. Understanding how solid state devices work and how they are combined to produce practical results will increase your skill and understanding of the subject. At intervals throughout the book, you will be able to apply what you have learned by building a "project." These "projects" are useful pieces of equipment which can be employed in these and other experiments. By the end of the book, you will have experienced the excitement of many phases of solid state electronics and will have actually built a useful experimental laboratory.

The Table of Contents lists the subjects of experimentation and project building. Each chapter covers one phase of solid state electronics and provides instructions for building one or more "projects."

Chapter 1 opens the door with ideas on how to get started in experimenting and project building with a discussion of materials and techniques. You start building with a simple power supply and a milliammeter/voltmeter.

Chapter 2 introduces solid state with the basics of various forms of diodes. Two useful projects are described; a germanium/silicon diode tester and an AC/RF ammeter.

With Chapter 3, transistors are introduced into the picture with theory and experiments. The project for this chapter is a transistor checker.

Power supplies are the subject of Chapter 4; both the regulated and unregulated varieties. Three power supply projects are described.

In Chapter 5 the subject is silicon controlled rectifiers and the project is to build your own light dimmer.

Chapter 6 deals with photo-sensitive devices. Projects include a light meter, an optical coupler, light beam communications and an optical fiber transducer.

Solar cells dominate Chapter 7. Theory and practice are discussed. The project is a solar battery powering a portable radio.

Field effect transistors are the basis of Chapter 8. The projects are an automatic level control module and a static sensitive light.

Building your own experimental laboratory is outlined in Chapter 9 with specific details of many useful boxes of circuit components. Two instruments are described as projects; a microvolt/milliammeter and a square-wave test generator.

Finally, Chapter 10 deals with linear integrated circuits with many combinations involving operational amplifiers. The climax of the project program of the book is to build a harmonic analyzer and a signal generator for the harmonic analyzer.

Upon completing the experiments and projects, you will have a greatly expanded understanding of solid state electronics and a laboratory of instruments of permanent value and usefulness.

A. Barber

Table of Contents

Experimenter's Guide to Solid State Electronics Projects

1

The Excitement of Experimentation and Project-Building

Experimenting and project-building involve one in the excitement of making new discoveries and of building and creating one's own electronic equipment. Understanding through experimentation heightens interest. This interest grows with each new insight into the mysteries of solid state electronics. Projects may be the climax of experimentation, giving the satisfaction that always comes with creation. You can even improve on the projects described in this book, or modify them to your own particular need. You may even surprise yourself with some new invention that fills an unsatisfied need. Solid state electronics is particularly adapted to experimentation, project-building and even inventing due to the small sized, inexpensive and easily procured components and because of the endless variety of combinations and functions that can be produced.

"Experimenting" means measuring and observing the electrical response and the light and heat generation of those tiny solid state electronic devices. It provides answers to questions such as: What bias voltage is safe? What current can the device carry without overheating? Does the device require a heat sink under proposed load conditions? Which device will cause the least distortion in my amplifier circuit? What basic circuit is best suited to my application? How much amplification can I obtain from this or a similar device? What variation in characteristics can be tolerated from device to device? How do various devices fail? How can devices be protected? How do various devices actually work to provide a given type of operation? How can I improve a particular circuit?

Experimenting accomplishes many ends.

1. *Experimenting aids in understanding the operation of components and combinations of components.*
2. *Experimenting shows the operating range of components in terms of voltage, current and temperature.*
3. *Experimenting helps find the best components or combinations for a given application.*
4. *Experimenting leads to improved operation of various projects.*
5. *Experimenting leads to new devices and even inventions.*
6. *Experimenting is often very important in exploring the practical value of a new invention.*

Projects are a natural adjunct to experimenting.

1. *Projects may be simple or complex instruments enhancing experimentation.*
2. *Projects give lasting satisfaction through repeated use.*
3. *Projects embody your understanding and become more meaningful than purchased equipment or even a kit.*
4. *Projects save money even when compared with most kits.*
5. *Projects grow by combining to extend the range of experimentation or utilization.*
6. *Projects give the builder a greater sense of accomplishment than kits can.*

7. *Projects are easier to repair since the builder knows all about the circuitry involved.*
8. *Projects can be a rewarding and interesting hobby, or they can help one's professional career.*

HOW I WAS TURNED-ON TO ELECTRONICS

Framingham, Massachusetts was a fertile field for a budding electronics experimenter in 1920. Our high school had an active radio club and after school anyone with even a crystal detector and a long antenna could hear the whine of rotary spark gap transmitters all over town. One evening one of the more advanced amateurs (Hams) invited me to his house to see his equipment. He let me listen to variously high pitched dots and dashes he said were coming from all over the world. I was turned on to a life of adventure in the world of electronics.

I had a crystal detector, a tuning coil and a long single-wire antenna. I could listen to local spark transmitters. I rigged up a Ford spark coil and could send and receive for a mile or so. I attended the Radio Club meetings and bought the few wireless magazines available at that time.

The **Boston Globe** ran a radio page every Saturday on which they described a new circuit every week. I needed a vacuum tube in the worst way to build these circuits. An audiotron, a two filament triode made for amateurs, was available at $6.00. I had earned the money, but my father said it was too much to spend on my hobby. I lied and told him that if he would let me buy this tube, I'd never ask to buy any wireless parts again. I rode my bicycle to Waltham, bought the audiotron and started building the interesting circuits published each week in the **Globe.** Typically, I would start each Saturday morning as soon as the paper arrived in the local store. I would not stop until I had a working device, which would often be Saturday night or early Sunday morning.

My commitment was complete when I picked up voice and music one evening from the Amrad station at Medford, Mass. I have been experimenting and building ever since. There is no end to what I have learned and to the excitement of creating new circuits, devices and systems which I have experienced and am still experiencing. I have been granted over 50 patents.

I hope that this book will bring some of my excitement to your life and will start you and encourage you to follow a life turned on by the wonder and mysteries of solid state electronics.

BEFORE YOU START EXPERIMENTING

There are certain basic things you must understand to get the most out of experimenting and project building. You must be able to understand a schematic circuit diagram. First of all, you must recognize the symbols used to represent the various circuit elements such as resistors, capacitors, transformers, rectifiers, transistors and so on. Figure 1-1 shows the symbols to be used in this book, together with their titles and a brief description of what

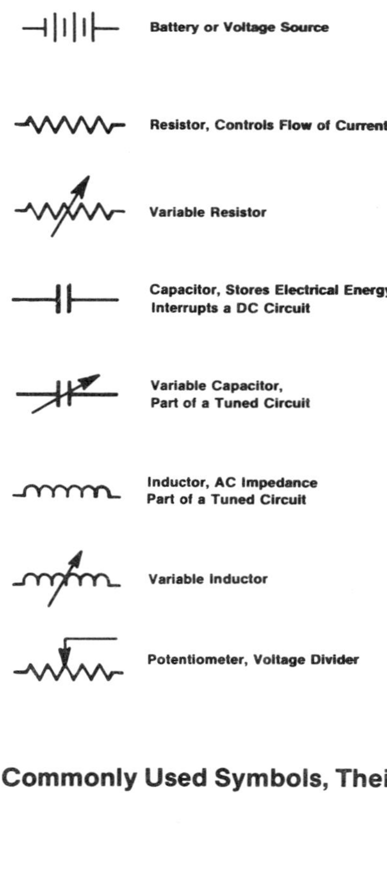

Figure 1-1—Most Commonly Used Symbols, Their Names and Uses

Iron Cored Inductor

AC Transformer, Steps AC Voltage Up or Down,
Audio Transformer, Changes Voltage or Impedance

Diode Rectifier, Changes AC or DC in Power Supplies, Radio Detector

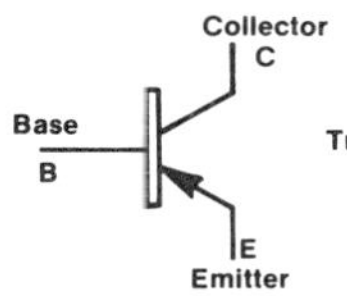

Transistor (PNP), Amplifier

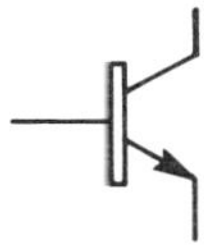

Transistor (NPN), Amplifier

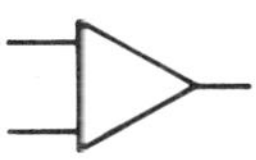

Amplifier, Such as an OP AMP (Operational Amplifier)

Meter, Measures I or E

each does. These elements, however, only become meaningful when combined into circuits. A **circuit** is one or more closed loops providing some desired result. Current flows and results are accomplished only when the loop is closed as shown in Figure 1-2, a simple circuit which demonstrates the basic law of all circuits, namely, Ohm's Law.

Ohm's Law can be written three ways, as shown in Figure 1-3. The first formula says you can measure R, if you know E, by

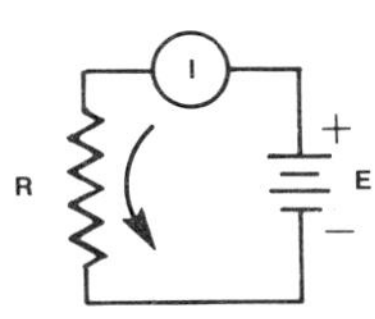

Figure 1-2—Simple Circuit

measuring the current I; or if you know R you can find E by measuring I; or, finally, you can find I if you know R and E. Thus, you have the basis for an ohmmeter, a voltmeter or a current meter. The arrow is added to indicate the direction of current flow, from the + (positive) battery terminal around the circuit to the − (negative) terminal (Figure 1-2).

$$R = E/I \qquad E = IR \qquad I = E/R$$

Figure 1-3—Ohm's Law

One can go a long way in experimental electronics with very little mathematics, although some people find mathematical manipulations may be very interesting or even fascinating. However, in any case, a few basic formulae should be learned and will prove to be great timesavers. In addition to Ohms Law, Kirchoff's two laws (See Figures 1-4 and 1-5) should be remembered. One of these laws states that at any junction point A (Figure 1-4), the sum of all currents entering and leaving that point is zero (logical since there can be no net accumulation of current at a point); and, the second law states that around a closed loop, all the voltages must add to zero (Figure 1-5), (again logical, since there can obviously be no voltage left over and unaccounted for). In these two equations the plus and minus values must exactly balance.

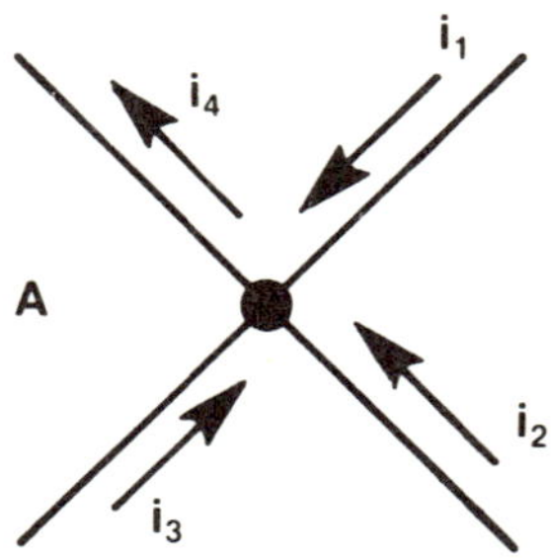

$$i_1 + i_2 + i_3 - i_4 = 0$$

Figure 1-4
Kirchoff's First Law

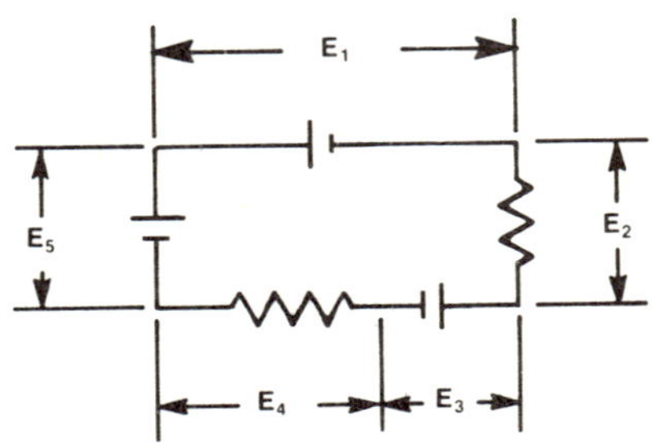

$$E_1 - E_5 - E_4 + E_3 - E_2 = 0$$

Figure 1-5
Kirchoff's Second Law

A little further mathematics, and you will be ready to solve many practical problems. Figure 1-6 shows two combinations of resistors. The two resistors in series simply add to the sum of the two. However, a more complicated situation involves two resistors in parallel. A series of resistors is needed when you need a resistor for which you do not have an accurate value, say for a voltmeter series resistor of 110,000 ohms. You do have them at 100,000 ohms and 10,000 ohms, and you make the 110,000 ohm resistor by connecting the two in series.

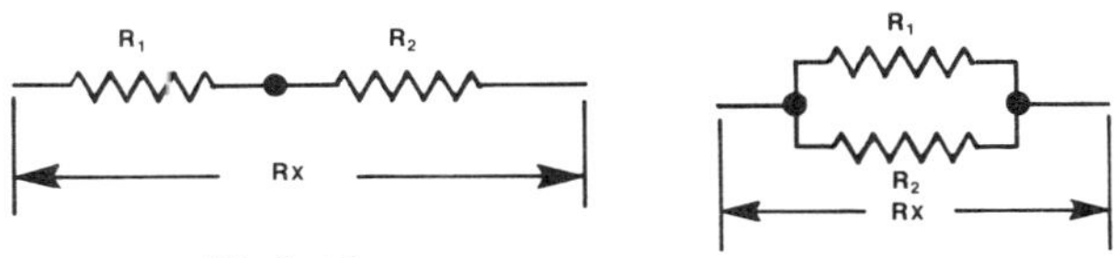

$$Rx = R_1 + R_2$$

$$\frac{1}{Rx} = \frac{1}{R_1} + \frac{1}{R_2}$$

$$Rx = \frac{R_1 R_2}{R_1 + R_2}$$

$$R_2 = \frac{R_1 Rx}{R_1 - Rx}$$

Figure 1-6—Resistor Combinations

Resistors in parallel come up most often when you want a shunt, when you are increasing the range of a milliammeter and you need a value you don't have. The desired resistor can be provided by connecting two resistors in parallel. Suppose, for example, you need a shunt of 80 ohms and the closest you have is 100 ohms. Call this R_1 and solve the third equation for R_2. It turns out to be 4000 ohms. Again if you don't have 4000 ohms, you can add two 2000 ohm resistors in series or take the nearest one to 4000 ohms and calculate the parallel resistor to produce 4000 ohms. For example, 4700 ohms requires 26,857 ohms in parallel, but perhaps you can find a more practical combination.

BASICS OF ASSEMBLY AND WIRING

A neat, carefully assembled and wired project gives lasting satisfaction to the experimenter. With the materials and parts available today this is not difficult. The subassemblies are conveniently made on pieces of perforated phenolic board called "perf" board. The board with the 0.1 × 0.1 hole spacing is most universally useful. One can lay out a circuit on such a board almost exactly as it is shown on the schematic diagram, particularly with solid state active components. With the parts mounted on the board (cut to convenient size) they are interconnected by the component's own leads or by adding pieces of copper wire. Where leads cross one another, insulated wire is required.

In order to complete a project, it is generally important to mount the subassembly in a bakelite box of proper size. Meters, switches and other controls are mounted on the top panel, thereby providing a finished product of lasting value.

Soldering Techniques

Soldering is not really difficult when properly learned. Just follow the directions below, and practice until you feel you have mastered the technique.

1. Employ a proper sized soldering iron. For most solid state projects, a small iron (30-40 watts) is adequate.

2. Use the proper tip on the iron. A chisel point ⅛ inch in diameter is recommended for the fine work which is most common with solid state circuits.

3. Keep the point well tinned and free from any oxidation products (black accumulations). A cleaning sponge designed for the purpose should be at hand and the tip should be wiped across the sponge at frequent intervals.

4. Always wait for the iron to reach operating temperature before attempting to solder. Usually about 2 minutes after plugging in.

5. Heat the joint to be soldered for a couple of seconds and then apply (rosin core) solder between the joint and the iron until the solder flows freely **around** *the joint. Don't keep iron and solder at the joint more than a couple of seconds. Remove iron and solder. When the solder solidifies, its appearance will change to a sort of silvery mat surface. Inspect the joint to see that the solder has wetted the joint wire, as shown by a slight flow along the wire. If not sure, reheat with soldering iron to produce a reflow. In any project, inspect all soldered joints with a magnifying glass. One poor joint can ruin the whole device.*

Mastering the art of soldering may take a little patience and practice at first, but is well worth the effort in the long run.

SETTING UP A PLACE TO EXPERIMENT

The experimenter needs some adequate space to set up and perform experiments, to do some mechanical work, to store parts and test equipment. A work bench or kitchen table will do to start together with some storage shelves or cabinets. A second bench or table is most useful for doing the mechanical assembly. This latter, for example, should be provided with a drill press, a jig saw, a vice and a soldering iron. A shelf above the first bench or table for the test equipment and a cabinet of small drawers for parts will complete a beginners accommodations.

Basic Tools for Project Building

Figure 1-7 is a photograph of a basic set of small tools for the experimenter. Most of these tools are available at any good hardware store with the possible exception of the wire strippers. The small wire strippers carried by most hardware stores can be used, but the professional type shown are so much more convenient and time-saving that they are well worth the extra cost.

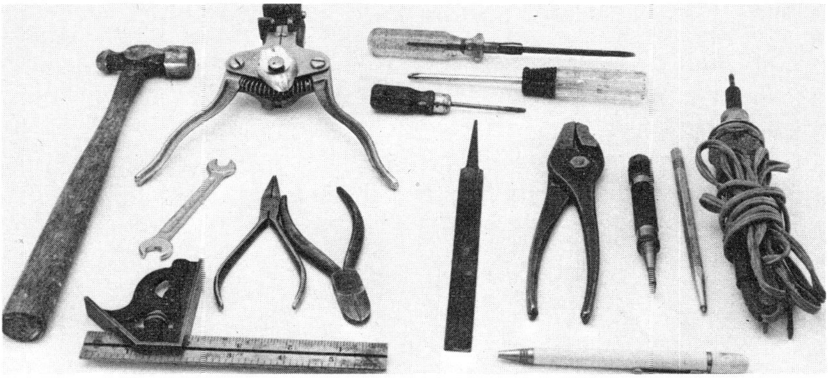

Figure 1-7—Small Hand Tools for Experimenting

Materials Most Involved

Following are the most commonly used materials used in making experimental set-ups. No. 24 bare wire is most useful for wiring on perforated boards. The insulated No. 24 stranded is most useful if several colors are at hand. Rosin core solder 50/50 or 60/40 and of minimum gauge is recommended. The perforated board should have 0.1 inch hole spacings. Again, these are the minimum amount of materials necessary and will, no doubt, be added to from time to time as experience dictates.

Experimenter's Test Equipment

This book assumes that the experimenter will have or buy some test equipment and will build more as described herein. Available test equipment is useful and important but there are many gaps which the experimenter can fill and at the same time get enjoyment and satisfaction in building for himself as well as saving money.

Start with a versatile multimeter such as the Radio Shack Cat. No. 28-4014. The experimenter can proceed, by following this book, to create his own test laboratory. Three major items which, though widely useful, have not been described are the cathode ray oscilloscope, audio generator and radio signal generator. He may wish to purchase these as the experiments and requirements progress. Relatively inexpensive equipment is generally satisfactory for experiments set forth in this book.

PROJECT: Experimenter's First Power Supply

Many experiments can be performed using a simple adjustable power supply. You can make this very useful power supply in an hour or so, using readily available parts at minimal cost. Figure 1-8 shows the circuit diagram including the line transformer, T_1, which

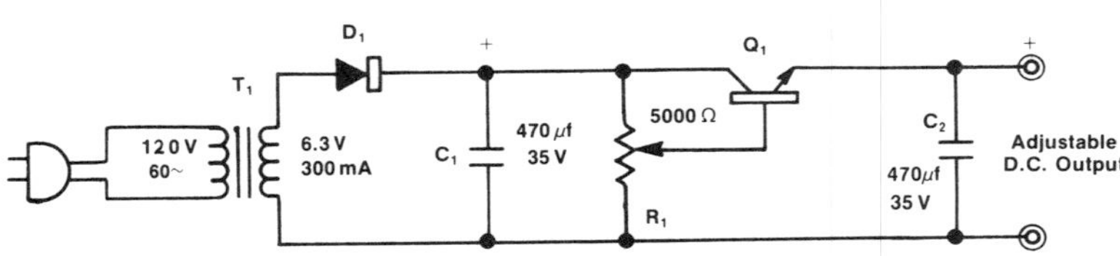

Figure 1-8—Power Supply Circuit Diagram

is called a step-down transformer, converting the 120 V. ac line voltage to 6.3 volts, the required level for the power supply; rectifier diode D_1, which conducts only on positive halves of the applied ac voltage and thereby charges capacitor C_1 to the peak voltage of this ac or $1.414 \times 6.3 = 8.9$ volts; connected between the positive side of capacitors C_1 and the positive output terminal or binding post is transistor Q_1, an NPN type, with its collector connected to capacitance, C_1, and its emitter connected to the + binding post; potentiometer R_1 is connected across capacitor C_1 and the variable contact indicated by the arrow is connected to the base or control element of transistor Q_1; a second or negative (−) output terminal or binding post is connected to one end of the transformer secondary and the negative side of C_1 and a second capacitor, C_2, is connected across the two binding posts. Transistor Q_1 acts as a variable resistor, controlled by the dc voltage applied from potentiometer R_1 to its base. The more positive this base voltage is made, the lower the resistance of transistor Q_1 and the higher the voltage across the output terminals. The output voltage can be controlled in this way from maximum voltage down to essentially zero. However, if no resistance or dc load is connected across the output terminals, capacitor C_2 will charge up and, with no discharge path, will remain charged.

In this particular project the following parts were used: (All were obtained from my local Radio Shack store.) Where **(part)** is used it indicates that the project used only part of the quantity purchased and is priced accordingly. Remaining parts can be accumulated for later projects.

Description	Radio Shack (Part No.)
4 × 2¼ × 2¼ bakelite box	270-231
Mini filament transformer	273-1384
5000 ohm potentiometer	271-1714
470 uf 35v. capacitor (2)	272-1018
Power transistor (NPN)	276-1634 **(part)**
Binding posts (2)	274-661 **(part)**
Potentiometer knob	274-1603 **(part)**
Line plug and wire	278-1255
Transistor mounting hardware	276-1371
Perfboard	276-1392 **(part)**

Power Supply Construction Details

The metal panel of the bakelite box is drilled in accordance with the drawing in Figure 1-9. The pattern for the transistor mounting is taken directly from the mica mounting washer. The mica washer is to be coated on both sides with silicon grease and

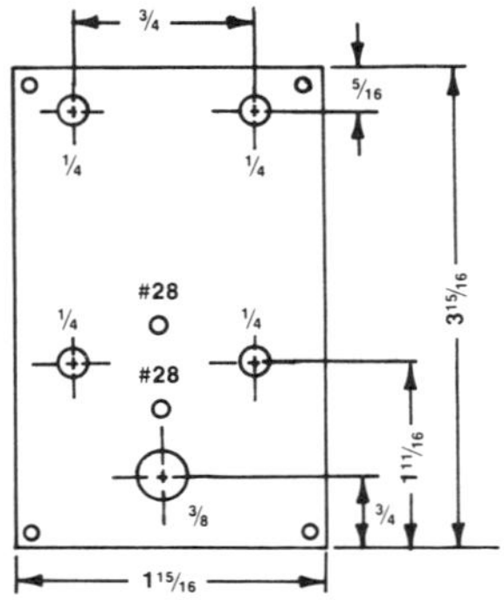

Figure 1-9—Drilling for Front Panel of Power Supply

the transistor is mounted through it on the metal panel. It is thus electrically insulated from the metal panel, which in turn acts as a heat sink. The transformer is mounted on the perf board with epoxy cement; it can also be mounted with screws and nuts. The transformer lugs should be cut a little short to allow fitting in the bakelite box. The capacitors are mounted by passing their leads through the perf board holes and bending over or returning through other holes for rigidity. Color coded wires are used for wiring, which is very simple, as can be seen from the drawing. The internal appearance is shown in Figure 1-10. The finished power supply is shown in the photo of Figure 1-11.

This simple power supply is unregulated, which means that the terminal voltage will drop slightly as the loading is increased. However, because it is adjustable, it has many uses in the field of experimentation.

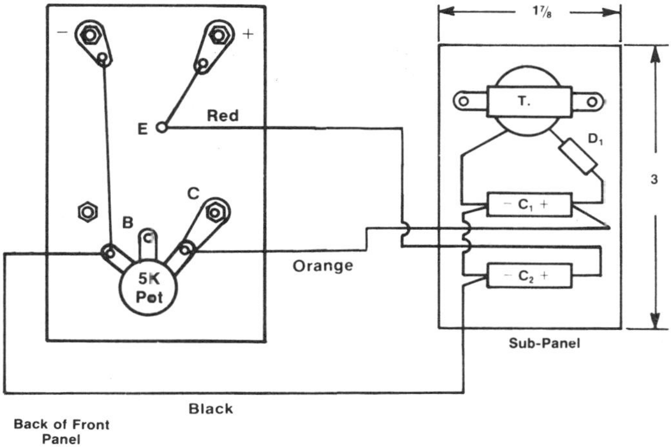

Figure 1-10—Inside the Power Supply

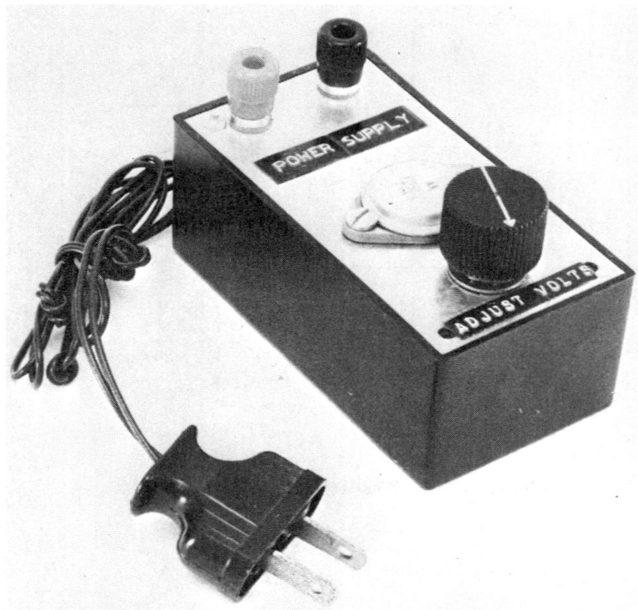

Figure 1-11—Completed Power Supply

PROJECT: Experimenter's Milliammeter/Voltmeter

To complement the power supply, the second proposed project is a combination milliammeter/voltmeter, which will provide the means to start experimenting. A milliammeter/voltmeter is a multirange instrument with a unique ability to measure both current and voltage in the same circuit by merely throwing a switch. The schematic circuit diagram is shown in Figure 1-12. Three current and three voltage ranges are provided. Assuming the meter is a 1 ma (milliampere or one-thousandth of an ampere) full scale, 200 ohm meter such as Lafayette catalogue no. 22-052, the ranges are 1, 10 and 100 ma and 1, 10 and 100 volts full scale. The first current position is the 1 ma range and no shunt is required. For the 10 ma range, the shunt is 22 ohms and for the 100 ma range, the shunt is 2 ohms. The series resistance for the 1-volt range is 800 ohms so that the 1 ma current through 200 plus 800 (or 1000) ohms is 1 volt; for the 10-volt range, 9800 ohms (the accuracy will not suffer in using a 10K resistor especially if one can be found that is 2% low) and for the 100-volt range 100K (K means 1000). The range-

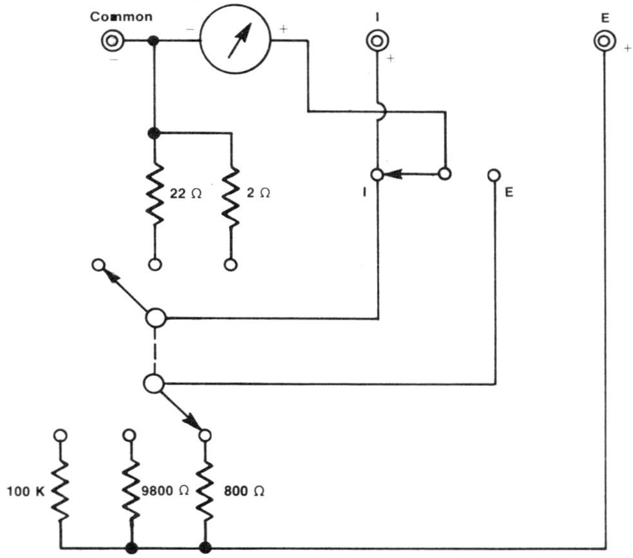

Figure 1-12—Milliammeter/Voltmeter Circuit Diagram

change switch is a 2-pole, 3-position **shorting** type. Shorting means that the variable contacts make contact with the stationary contact of a succeeding position before opening contact with the preceeding position. This is important in a current-range switch since each time the range is changed, the meter will be unshunted during the transition, possibly damaging the meter movement. The three terminals may be binding posts or pin tip jacks. I prefer the latter as will be clear from a later chapter. The I to E changeover switch is a mini SPDT (single-pole double-throw) toggle switch.

The parts used in my model were as follows:

Description	Lafayette Part No.
2″ 1 ma. meter	99A51070
Pin tip jacks (3)	32P64900
Rotary switch	99A64206
Toggle switch SPDT	34A02328

(cont'd) Description	Lafayette Part No.
½ watt resistors, one each 2, 22, 800, 9800 and 100K	
Phenolic box 6 × 3½ × 1⅞	Calectro Cat. No. H4-726

Front panel drilling is shown in Figure 1-13. Wiring was carried out without a sub-assembly board as shown in Figure 1-14. The completed unit is shown in the photograph Figure 1-15.

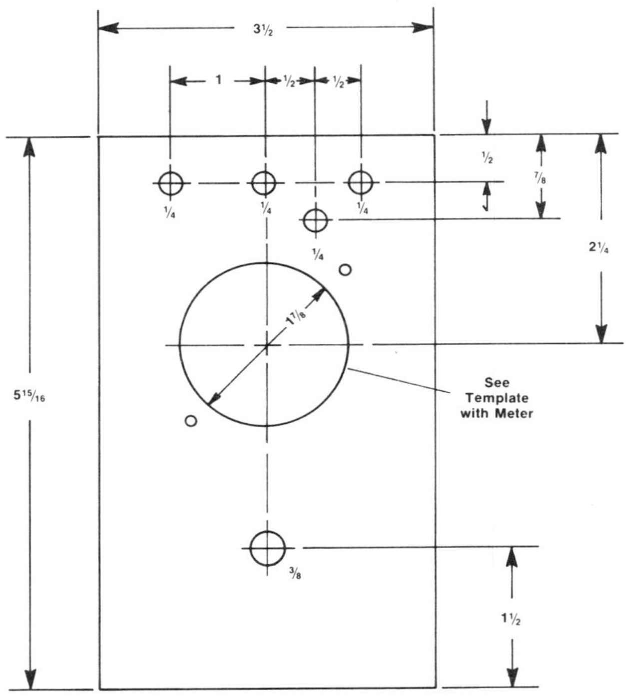

Figure 1-13—Panel Drilling

Example of the Use of the First Two Projects

The milliammeter/voltmeter has a very convenient and unique capability. With a common terminal for both current and voltage

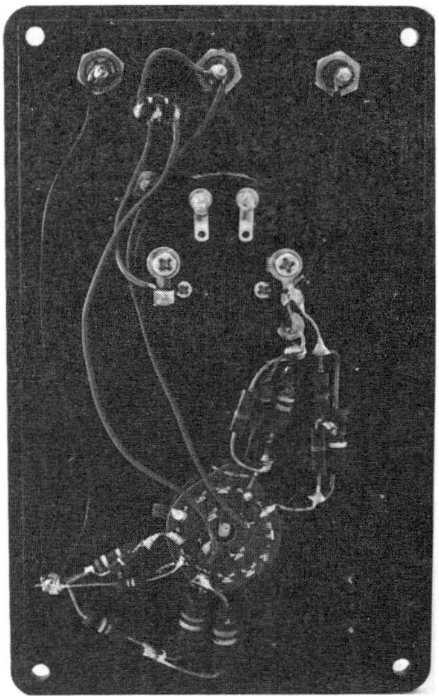

Figure 1-14—Milliammeter/Voltmeter Assembly and Wiring

Figure 1-15—Milliammeter/Voltmeter Complete

measurements, many set-ups can be made where current is observed and voltage is checked by merely throwing the I/E switch without changing circuit connections. An example is shown in Figure 1-16 where the h_{FE} or current gain of a transistor is being measured.

At any rate, if you have completed the first two projects, you have an adjustable power supply and a versatile milliammeter/voltmeter which can be used in many experiments outlined in later chapters.

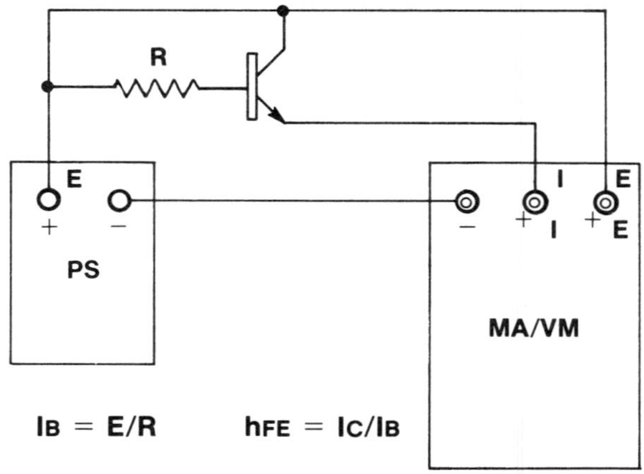

Figure 1-16—Set-Up for Measuring Current and Voltage

2

Projects and Experiments with Crystal Detectors, Diodes and Rectifiers

Materials used in electrical and electronic circuits and devices are grouped into three broad general classes: conductors, semiconductors and insulators. Electrical conduction or current flow is primarily the flow of tiny particles called electrons. These particles are found in the atom where they orbit around a nucleus, much as the earth orbits around the sun. We are mainly concerned with solid materials that contain many electrons in each atom. Each electron represents a tiny charge of negative electricity, and in a typical atom, the nucleus represents a positive charge equal to the sum of the negative charges of its orbiting electrons so that all charges are normally neutralized.

When an electrical current flows, some of these electrons are pushed or pulled from their orbits and moved to new orbits around other nuclei, being replaced by new electrons. In materials where electrons are easily displaced from their orbits, electrical current flows readily and the material is classified as a conductor of electricity. All metals are conductors of electricity in varying degrees. Silver is the best conductor, while copper is next best.

Copper and aluminum are generally used for electrical conductors, although in special situations other metals such as silver and gold may be used.

When the electrons are very tightly bound in their orbits, the material becomes a non-conductor or insulator of electricity. Glass is a classic example of an insulator. Practically no current flows through a piece of glass even when 1000 volts or more is applied across it.

Between the two materials classed as conductors and insulators there is a class called semiconductors. In these materials, the electrons are bound more tightly than in conductors, but not as tightly as in insulators. They conduct electricity to a limited degree, particularly when they are in a pure state. However, impurities can be added to semiconductor materials which greatly alter their conductivity. These semiconductors, generally with added impurities combined in various ways, provide the basis for solid state electronics.

AC AND DC CURRENTS

Solid state electronics deals with two types of electrical currents, namely direct current (dc) and alternating current (ac). Direct current is the steady current, for example, the current provided by a battery. As for a battery, it has a positive (+) and a negative (−) polarity with respect to a given point in the circuit. The reference point is often called the common point or ground (G).

Alternating current on the other hand is a constantly changing current. It oscillates about its reference point; part of the time it is positive and the rest of the time it is negative. Like the agitator in a washing machine, it moves in one direction for a while and then reverses and moves in the other direction. A typical example is the sine wave, a smooth up and down oscillation as shown in Figure 2-1. Conventionally, the portion above the center or common line is called positive while that below the line is called negative.

Alternating current is the type of current supplied over power lines; it is the current driving the loud speaker of a radio; and the carrier wave which carries radio and television signals through the air. Much of solid state electronics is concerned with converting alternating current to direct current and vice versa.

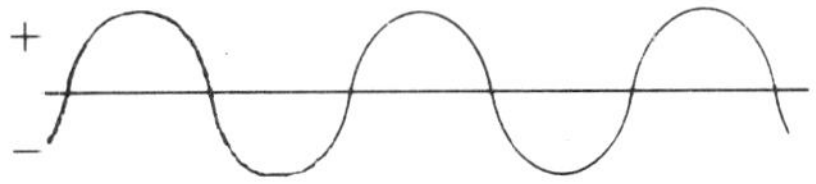

Figure 2-1—Sine Wave

PN DIODE/RECTIFIERS

The most widely used semiconductor base materials are germanium and silicon. Each, in its pure state, is a relatively poor conductor of electricity and has little usefulness. However, if certain other elements are added, they not only become much better conductors, but assume important characteristics. If arsenic is added to pure germanium in a minute quantity it becomes more conductive due to a greatly increased number of loosely held electrons in the resulting material. It is said to have excess electrons and is called N material. These electrons can be drawn from the N material by a positive potential, since unlike polarities attract just as unlike poles of a magnet attract. However, as will be apparent from what follows, merely making the silicon or germanium more conductive by adding impurities has no particular use in itself.

If gallium is added to pure germanium or silicon, an entirely different effect is produced. While there is no such thing as an electron carrying a positive charge, this "doping," as it is called, makes the semiconductor act as though it had an excess of loosely held positive charge carriers. Actually, this impurity has too few electrons to make a neutral combination, resulting in an apparent excess of positive charges. This is called positive or P material. Again, this characteristic in itself has little use.

The interesting and very significant fact is that if a piece of N material is fused (grown) to a piece of P material, a very useful device results, known as a PN junction. Figure 2-2 is a fantasized picture of the PN junction or diode, greatly enlarged for purposes of explanation. The N portion is represented as positive nuclei surrounded by negative charges (electrons) while the P portion has negative nuclei surrounded by positive charges.

Now, if the P end is made positive with a positive potential as from a battery, the negatively charged electrons are attracted and flow across the junction, the dividing line between P and N, and a current flows in the circuit. However, if the P end is made negative, the negative charges carried by the electrons are repelled and no current flows. This, then, is a rectifier, a device which conducts in one direction and not in the other. It is called a PN diode or rectifier.

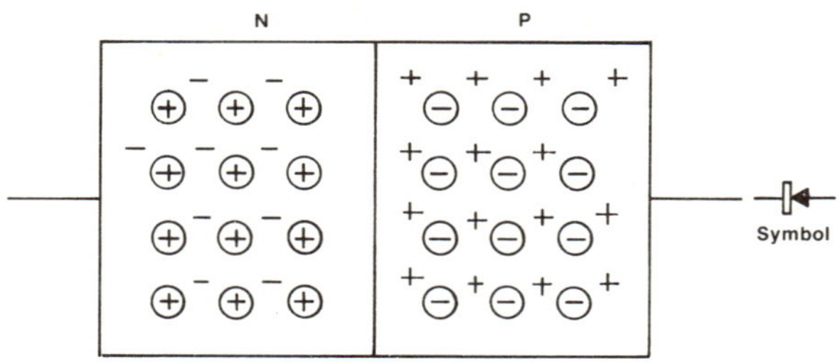

Figure 2-2—PN Junction

MEASURING DIODE CHARACTERISTICS

Diode characteristics are generally measured by observing the current flowing through the diode as a function of the voltage across it. The important parts of these characteristics are at low current and low voltage, at milliamperes and fractions of a volt, for example. The power supply used with the milliammeter/voltmeter which were Projects 1 and 2 of Chapter 1 makes a good combination for measuring diode characteristics. A set-up for doing this is shown in Figure 2-3.

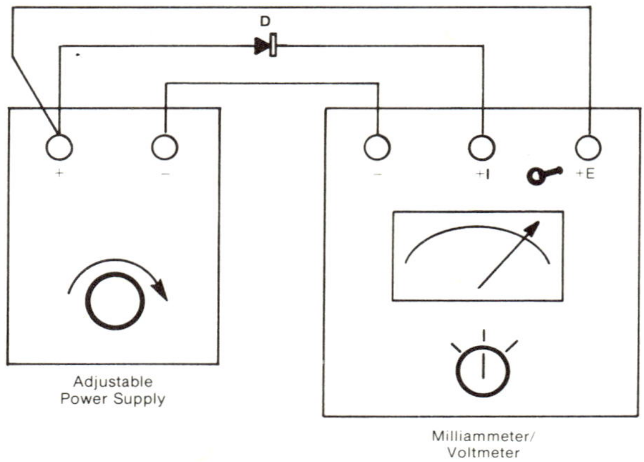

Figure 2-3—Set-Up for Measuring Diode Characteristics

In any circuit for measuring current and voltage simultaneously, at least one correction must be made for a high degree of accuracy. The current through the device being tested must be corrected for the shunt current flowing through the voltmeter, or the voltage must be corrected for the voltage drop in the milliammeter. See Figures 2-4 and 2-5.

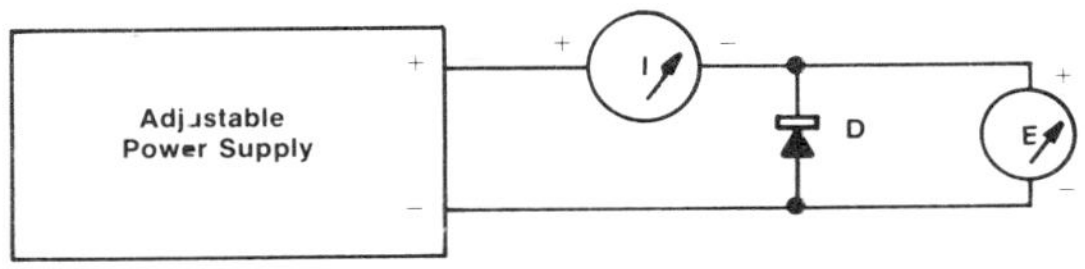

Figure 2-4—Circuit for Measuring Voltage and Current Simultaneously

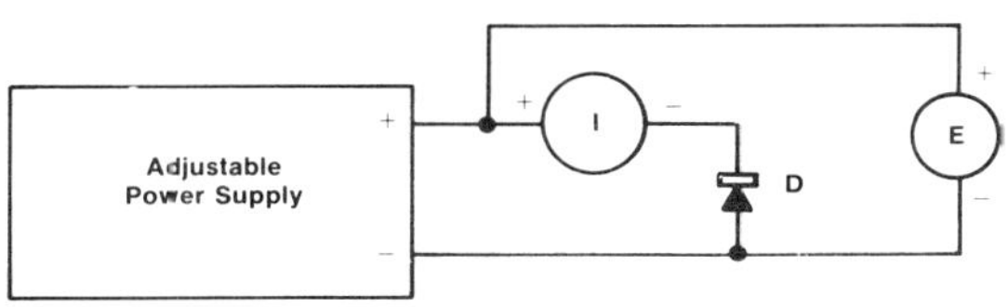

Figure 2-5—Another Circuit for Simultaneous Measurement

In Figure 2-4, the voltmeter is connected across the device under test, thereby measuring the voltage drop correctly. But, the voltmeter current flows through the current meter, requiring a correction. In Figure 2-5, the voltmeter is connected across both the current meter and the device under test, so that the current reading is correct. However, the voltage drop through the milliammeter must be subtracted to obtain a correct voltage reading.

It might be assumed at first glance that with a meter of a given basic sensitivity, on which both current and voltmeters are based, that it would make little difference which correction is chosen. However, as the current sensitivity of the voltmeter is increased, the circuit of Figure 2-4 becomes more desirable, since it will require less and less correction. For example, if one is measuring a current of 1 ma through a device having a resistance of 1000 ohms, the voltage drop across it should be 1 volt. However, a voltmeter having a 1 ma basic meter sensitivity will carry an equal current and the current meter will read 2 ma instead of the correct 1 ma. However, if the basic meter sensitivity of the voltmeter is 50 microamperes, it will have an impedance at 1 volt-full scale of 20,000 ohms and will cause the current meter to read 1.00005, an error of only .005 percent.

Turning to Figure 2-5 with the same set of conditions, what is the correction? There is a basic voltage drop required to deflect the current meter, no matter what the shunted sensitivity. This means that if the meter has a basic sensitivity of 1 ma, it may have a 0.2 volt drop at full scale. When shunted to read 100 ma, it will still show a 0.2 volt drop across its terminals. Going to a meter having a current meter sensitivity of 50 microamperes may make matters worse, since this meter may require a voltage drop of 0.25 volts to read full scale. Again, shunting the meter for higher current ranges does not change the drop across the meter, shunted or unshunted.

The meaning of all of this is, simply, that a basically sensitive meter can be a big advantage when used as a voltmeter, but it is probably no better, and may be worse for certain kinds of measurements when used as a current meter.

Measuring Diodes with the MA/VM Meter

The general discussion above is to be kept in mind when measuring diode characteristics, since their critical region is at low voltage and low current. Assuming, of course, that one has a choice of meters to work with.

In case the experimenter has only the power supply and milliammeter/voltmeter (MA/VM) of Projects 1 and 2 of Chapter 1, he will have to use the set-up shown in Figure 2-3 for measuring diode characteristics as outlined in this chapter. Here, the voltage and current are measured by flipping the I/E switch. When the current measurement is made, the voltmeter is disconnected by the switch and hence does not affect the current reading. When the voltage measurements are made, the current meter shunts are across the voltage source thereby loading the source (power supply) as it was in making the current measurements. No corrections are required in either mode. The following are some actual observations.

Range	Measured E	Measured I
1 ma	0.1 volt	0.1 ma
10 ma	0.5 volt	6.0 ma
100 ma	1.0 volt	34 ma

DIODES HAVE MANY USES

Besides the more conventional uses of diodes such as ac to dc power converters and modulated signal detectors, diodes have many other uses. Since these other uses often come in handy, a few will be reviewed at this time.

First, the one way conduction characteristic is useful for protecting sensitive equipment against reverse or incorrect polarity of an applied source of voltage.

An application along these lines is the diode which is used in a solar cell battery charger. Figure 2-6 is a simple solar cell charging circuit. The series-connected solar cells provide positive voltages at their anodes which add to provide sufficient voltage to more than equal the battery voltage plus the drop in diode D, and hence provide a source of voltage for charging the battery. However, it can be seen that, should the sunlight fail, the solar cell voltage will drop and the battery will discharge through the solar cells. The diode permits the forward-charging current to flow from the solar cells to the battery, but blocks any reverse flow from the battery to the solar cells.

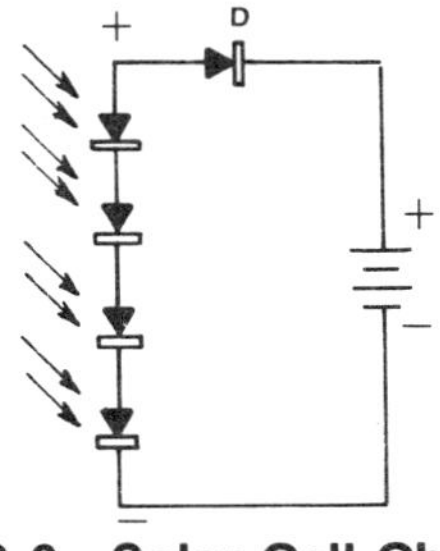

Figure 2-6—Solar Cell Charging Circuit

Second, diodes are used to absorb inductive kick-back voltage from pulse-operated inductive devices, such as magnetic relays.

Figure 2-7 shows a transistor driving a magnetic relay. The relay may be turned on by a pulse applied to the base of the transistor. Now, if the base signal calls for turning off the relay by a sudden drop to zero, the current supplied by the transistor also goes to zero. However, due to the current that had been passing through the relay coil (inductance), it has built up a large amount of stored energy. This energy induces a voltage of opposite polarity to the current which stored it, but the transistor has been turned off and cannot dissipate this energy. Consequently, in attempting to dump this energy, the coil builds up a high-voltage surge. Such a surge is very likely to damage the transistor. The diode, connected as shown will dissipate the inductive stored energy, while simultaneously limiting the back voltage applied to the transistor. The only requirement is that the diode be of sufficient current-carrying capacity to safely handle the initial current surge.

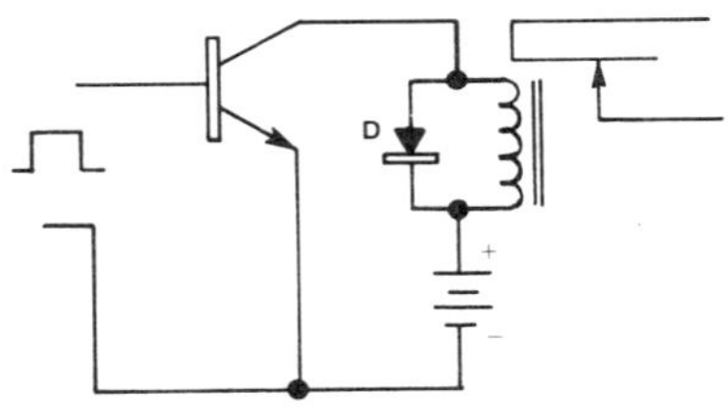

Figure 2-7—Transistor Driving a Magnetic Relay

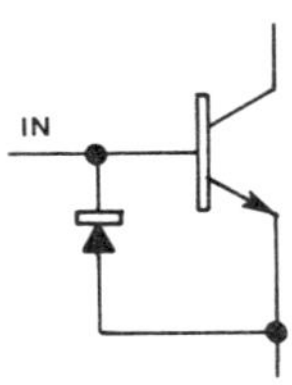

Figure 2-8—Diode Limiting Negative Direction Voltage

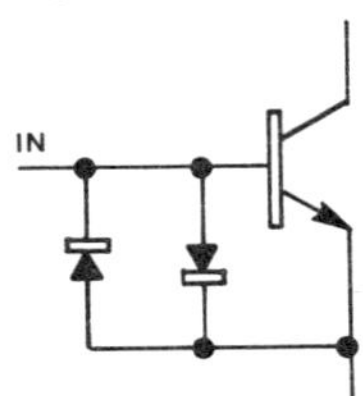

Figure 2-9—Limiting Configuration with Two Diodes

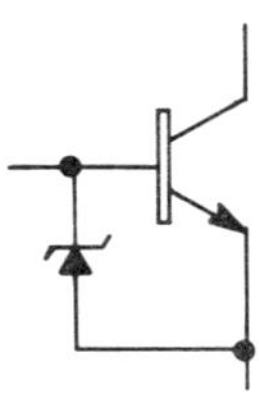

Figure 2-10—Circuit Using Zener Diode to Protect Base

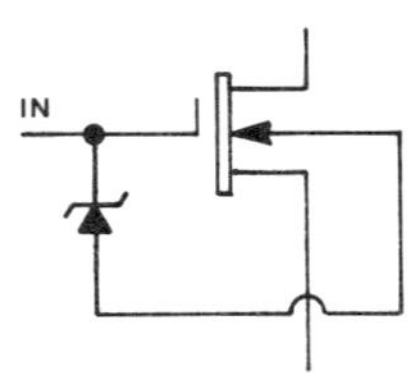

Figure 2-11—Zener Diode Protecting MOSFET Transistor

Third, diodes are used to protect transistor inputs. In Figure 2-8, a diode is connected between the base and emitter of a transistor to limit any negative direction voltage that might appear in the base input lead. In Figure 2-9, two reverse-connected diodes are connected from base to emitter. Such a connection, however, will limit forward bias on the transistor to too low a value to be practical. Figure 2-10 shows a circuit using a single zener diode for protecting the base. Here, reverse voltage is clamped at about 0.6 volt, the conduction voltage of a silicon diode, while the forward voltage is clamped at the zener voltage, which can readily be chosen to permit full forward bias on the transistor base before limiting. Figure 2-11 shows the same zener diode protection applied to an n-channel depletion type MOSFET transistor.

Diode protection is a natural for MOSFET transistors since the gate of such a transistor is extremely sensitive to overvoltage and, particularly, to static-induced voltages. A great convenience is provided when the diode protection is built into the transistor. Not only is the protection provided in an actual circuit, but it is provided prior to connection in a circuit so that handling these vulnerable transistors is made much easier and safer.

Diodes have still further uses. Diodes are used in logic gates called diode gates. A diode connected from output to input of an operational amplifier provides an amplifier with a logarithmic response. Many special purpose diodes having characteristics adapted to unique types of responses are also available.

DIODE PROJECT-PROTECTING METERS

Burning out a meter can be a sad experience for an experimenter. This can be especially sad if the meter happens to be in a piece of equipment which the experimenter has labored to build. Meter burn-outs occur most frequently when a sensitive meter is being used and a sudden surge takes place in the circuit, or when a voltmeter is connected across a circuit carrying a much higher voltage than the maximum of the range to which the voltmeter is switched.

A common method of protecting a sensitive meter is to connect a pair of cross-connected diodes across it. However, it is not as easy as it looks, simply because diodes have different break-down voltages and meters have different voltage drops across them in their working range. The protective diode must permit normal operation of the meter over its full range, but then clamp higher voltages as close to the maximum working voltage as possible.

The meters used in this book are what is known as D'Arsonval type instruments. They consist of a coil of fine wire with a pointer attached and pivoted in a strong magnetic field supplied by a permanent magnet. When an electric current flows in the coil, its lines of force interact with the lines of force of the permanent magnet, causing the coil to rotate and deflecting the pointer. Since it is current in the coil that causes the deflection, all such meters are really current indicating meters. They read voltage merely as the product of the current and the resistance of the meter and its multiplier resistors.

A typical 1 ma meter has a coil resistance of 200 ohms, and hence deflects full scale when $200 \times 0.001 = .2$ volts is applied to

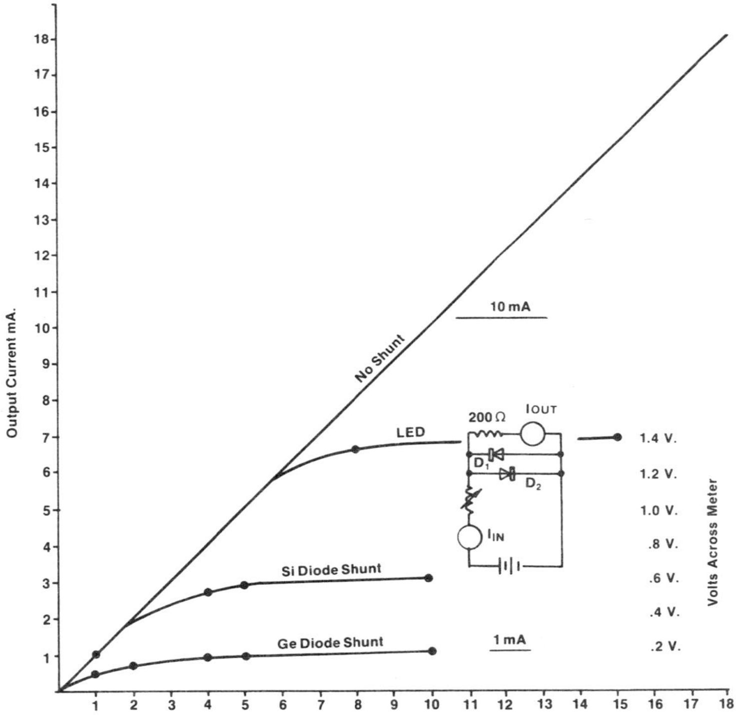

Figure 2-12—D-Arsonval Instrument and Graph of Limiting Characteristics

its terminals. To make a 1-volt full scale voltmeter merely add 800 ohms in series so that $1000 \times 0.001 = 1$ volt.

Figure 2-12 shows a meter having an internal resistance of R protected by two cross-connected diodes D_1 and D_2. Taking the figures above, if R = 200 ohms and the meter current sensitivity is 1 ma, the protective diodes should allow 0.2 volt across the meter but should clamp any higher voltage as sharply as possible.

Figure 2-12 shows a test set-up for evaluating the effectiveness of germanium, silicon and LED diodes (LED means light emitting diode) in protecting a 1 ma meter having an internal resistance of 200 ohms. The graph shows the results. The germanium diode diverts substantial current below 1 ma and so would not be practical. The silicon diode does not affect the simulated meter current of 1 ma but limits very sharply at 3 ma. This should be good protection against burn-out for a 200 ohm, 1 ma meter. The LED diode limits at about 7 ma which would probably still prevent burn-out of a 1 ma meter, but would not provide the safety of the silicon diode. The LED, however, has a unique advantage as a protective diode in that it gives a visual indication on severe overload.

In addition to the current scales, a voltage scale is shown. For example, the LED diode limits at 1.4 volts. In a later project two back-to-back LED diodes are used to protect VU meters which are rated at 0.8 volts ac at 0dB. This is 1.13 volts peak. The graph of Figure 2-12 shows that LED limiting starts at 1.2 volts dc so that the VU meter reading will not be affected up to 0dB reading but will limit rapidly above 1.2 volts. Thus, the LED diodes are ideal limiters for this particular VU meter. Furthermore, the LEDs give visual indication of overload, which also instantaneously gives good peak indication.

RECTIFICATION WITH A PN DIODE

If, instead of a steady voltage from a battery, an alternating voltage is applied to the PN diode, a unidirectional current pulsating direct current, or dc, is obtained. Figure 2-13 shows how when an alternating current sine wave voltage is applied to a PN diode a unidirectional pulsating positive output voltage is obtained as shown in Figure 2-14. In order to make this dc useful, the pulses must be smoothed out. This is done with a filter capacitor as shown

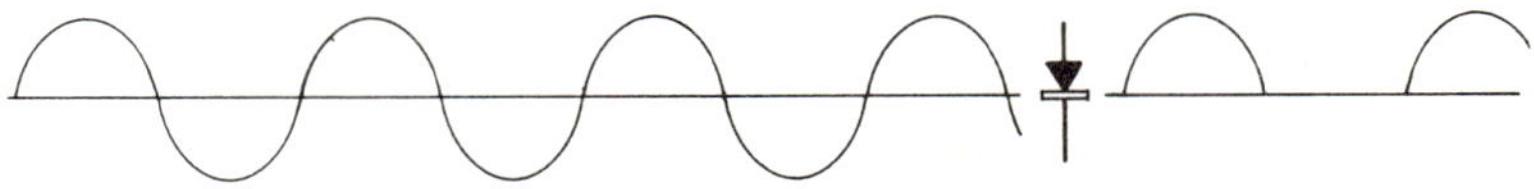

Figure 2-13—How a PN Diode Affects a Sine Wave

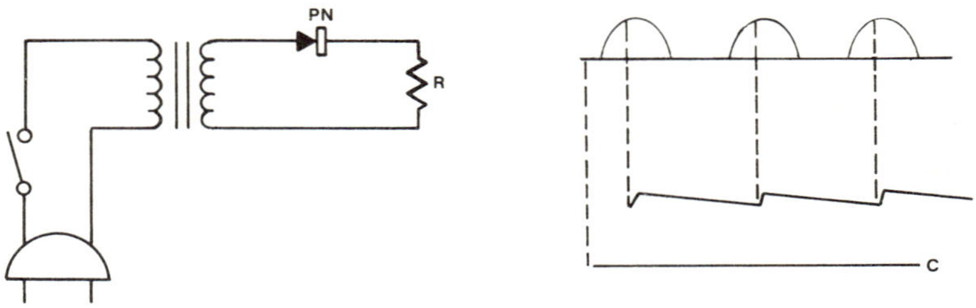

Figure 2-14—Circuit Diagram for PN Diode Rectifier with Resultant Waveform

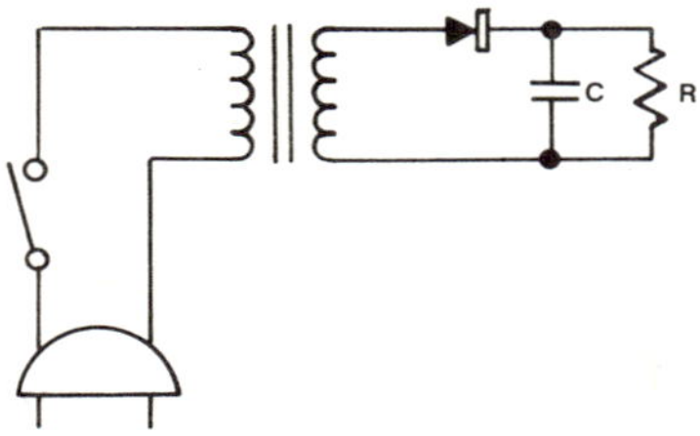

Figure 2-15—Circuit Diagram Adding Filter Capacitor

in Figure 2-15. The capacitor is charged through the rectifier whenever the alternating current peaks exceed the voltage across the capacitor and, in return, the capacitor discharges through the load represented by resistor R, as in Figure 2-15. The source of alternating current is shown as a transformer, with its primary to be plugged into an ac wall socket and its secondary connected to the rectifier, capacitor and load circuit. In solid state electronics, this transformer will generally be a step-down transformer, having a secondary voltage less than its primary voltage. A common ratio is 120 volts (primary) to 6.3 volts (secondary).

Signal Rectifier/Detectors

One of the earliest, yet most important applications of diode rectifiers is that of signal detection. The old crystal detector was a diode and the most successful of the early radio signal detectors. The detection of radio signals is slightly different from the production of dc from ac process used in power supplies. It is called demodulation.

Amplitude modulated (AM) radio signals consists of a high frequency carrier which is modulated in amplitude by the audio signals being conveyed. Figure 2-16 shows the high frequency alternating current carrier, the low frequency audio signal and the carrier modulated by the audio signal. This method of transmission is used, because the high frequency signal is easily sent out by a transmitter and travels with low loss in signal over great distances. An equivalent transmission of the audio signal without the carrier and without the use of wire lines as in telephone communication is most difficult. The function of the detector is to demodulate the modulated carrier in order to recapture the audio signals.

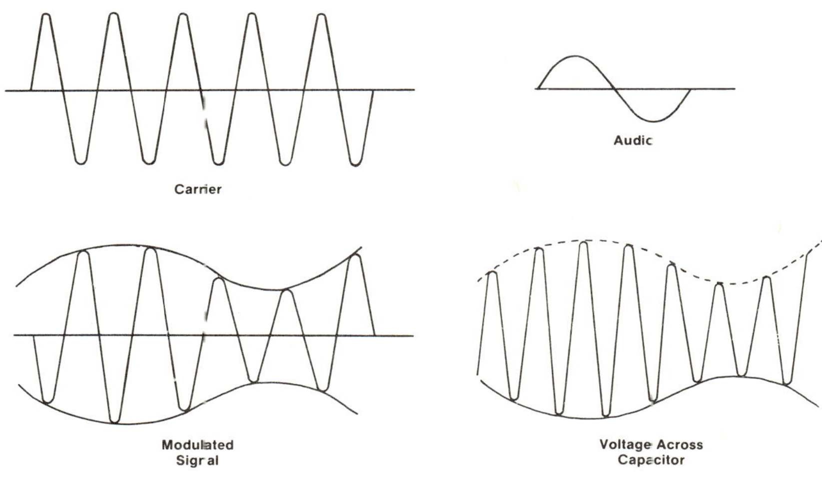

Figure 2-16—Carrier Modulation with Audio Signal

Figure 2-17 shows a simple, basic radio-type diode detector circuit. An input consisting of a modulated carrier is introduced through a radio frequency transformer and the secondary voltage of this transformer is applied to a diode detector (rectifier) and capacitor C_1 connected in series. Across the capacitor is connected a resistor R_1. Resistor R_2 and capacitor C_2 are also connected in series across the capacitor to filter out the carrier.

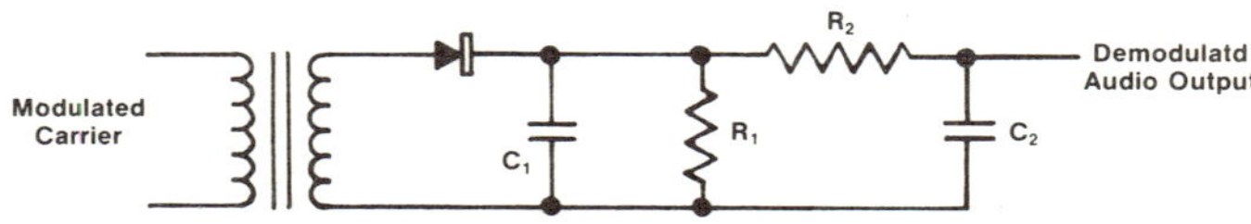

Figure 2-17—Diode Detector Circuit

The audio output derived from the demodulation process is taken from the junction between capacitor C_2 and resistor R_2. This is a basic circuit for performing the functions necessary to recover the audio frequency signal from the high frequency-modulated wave.

As in the case of the power supply, the ac of the modulated carrier passes in only one direction through the rectifier and charges capacitor C_1. However, since we are not interested in a dc output, the capacitor cannot be too large. The voltage across the capacitor must follow the modulations on the carrier, but not the carrier itself. Each time a peak occurs in the input carrier, the capacitor is charged through the rectifier to substantially that peak voltage. However, the next carrier peak may be greater or less than the first. If it is greater, there is no problem. The capacitor is charged to the value of the new peak. However, the next peak may be less than the first. In this case the capacitor must have discharged to a value not greater than the new peak or the resulting audio will not be a true reproduction of the modulation. Thus, the capacitor and its shunt resistor R_1 must be able to follow the modulation. If it does not follow, the capacity of C_1 or the resistance of R_1 must be reduced until it does follow. This naturally should account for the highest audio frequency and the deepest modulation for that frequency to be reproduced from the modulated wave. Resistor R_2 and capacitor C_2 are merely an R-C filter to remove carrier ripple from the demodulated audio signal.

The experiments indicated here are to determine C_1 and R_1 for a given carrier frequency, audio modulating frequency and depth of modulation. Modulation depth of percent is merely the ratio of the modulating wave amplitude to the carrier amplitude. Find out the effects of C_1 being too small or too large. Do the same thing for R_1.

The non-linearity of the diode detector and, especially, that of the silicon diode can be overcome to some degree by the use of a bias in series with the rectifier. A circuit for doing this is shown in Figure 2-18. The silicon diode starts to conduct at about 0.5 volt as shown in Figure 2-19. The bias approach has the greatest effect on signals having a high modulation percentage. For these, the flattening is substantially reduced and the demodulated signal amplitude is increased. The waveforms shown in Figure 2-20 illustrate the demodulated sine wave without bias, (A), and then with bias, (B), showing the improvements in waveform and amplitude.

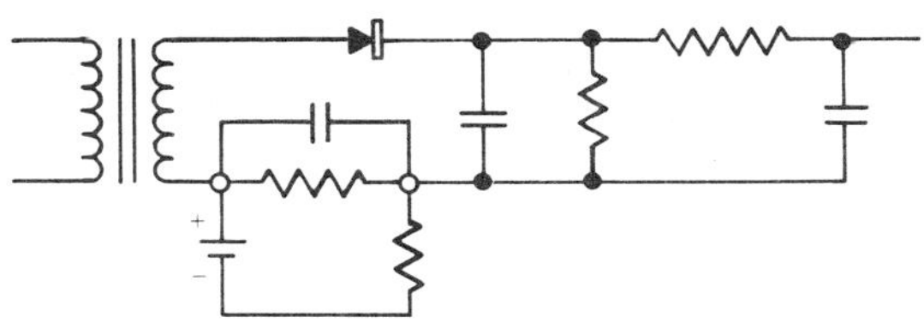

Figure 2-18—Bias in Series with a Rectifier

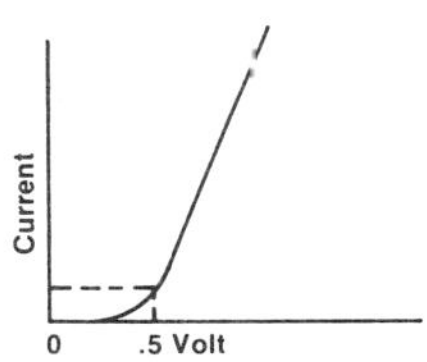

Figure 2-19—Silicon Diode Conduction

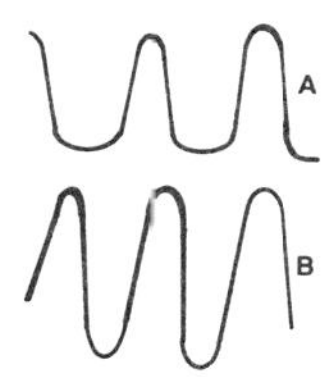

Figure 2-20—Sine Wave With and Without Bias

PROJECT: Ge/Si Checker

It is important to know whether a particular diode is germanium (Ge) or silicon (Si). Most of these tiny devices are unmarked and so cannot be identified. This situation makes a germanium/silicon checker most useful. The project to be described is simple and easy to build and should cost no more than about $10.00.

The checker is based on the fact that the forward voltage drop at low current is about 0.2 volt for germanium and about 0.6 volt for silicon. Thus, when a diode is shunted across a divider circuit between a battery and a sensitive meter, the voltage meter without the diode connected is multiplied to read 1 volt full scale. When a diode is connected across the multiplier path at the 0.5 volt drop point, the meter reads the anode to cathode voltage of the diode. Open or shorted diodes can also be detected. An open or reverse-connected diode will not affect the meter reading (1 volt). A shorted diode will drop the meter reading to zero.

One of the most helpful uses of the Ge/Si checker is in identifying the anode end of a diode.

Figure 2-21 is the schematic circuit diagram of the checker. The meter used in this particular project is a Radio Shack Cat. No. 22-0520-1 ma meter. The on/off switch is a push-on/push-off push-button switch, Radio Shack Cat. No. 275-011. The connections for the diode under test are provided by two pin tip jacks fitted with pin plug clips which plug into the jacks and are designed to receive wire leads. The resistors are half-watt carbon resistors. The 500 ohm resistor is provided by paralleling two 1000 ohm resistors and the 800 ohm is made by connecting a 470 and a 330 resistor in series. The meter resistance is 200 ohms.

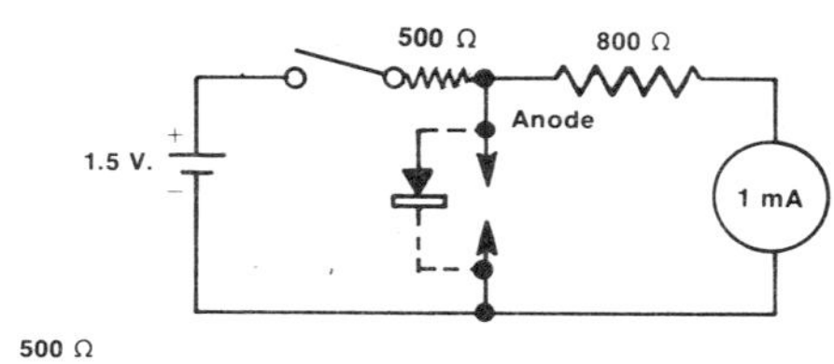

Figure 2-21—Schematic Diagram of Ge/Si Checker

Figure 2-22 is a dimensioned drawing showing a top view of the panel drilling, Figure 2-23, a photograph of the internal assembly and wiring and a photograph of the finished instrument, Figure 2-24, are also included.

To use the Ge/Si checker, merely push the push-button switch to ON. The meter should read full scale (1 volt). Clip in the diode to be checked with the anode to the right. A germanium diode will read approximately 0.2 volt and a silicon diode will read approximately 0.6 volt. Small variations one way or the other are not significant.

To identify the anode of an unknown diode, clip across the test clips. If the meter drops to 0.6 or 0.2 volts, the anode is to the right. If the meter reading does not change, reverse the diode. If it still does not affect the meter reading, the diode is open circuited. If the meter reads zero, the diode is shorted.

Turn the push-button switch to OFF when not in use. Since a battery is subject to very little drain, a fresh battery should last a year or more. The battery in the model was soldered in place since it will

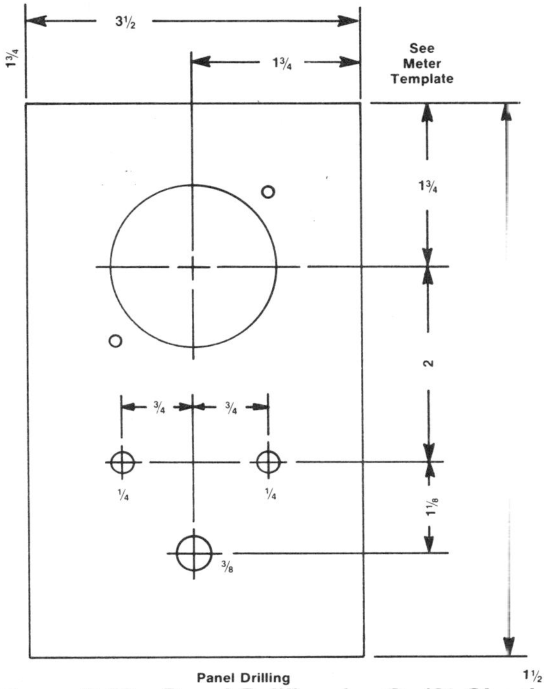

Figure 2-22—Panel Drilling for Ge/Si Checker

Figure 2-23—Ge/Si Checker Under Panel

Figure 2-24—Ge/Si Checker Complete Unit

not have to be changed very often. Of course, one could use a battery clip for the battery.

THE ACTIVE DETECTOR

One significant development in the art of detection is the active detector. This device takes advantage of the unique properties of the operational amplifier. The operational amplifier is a high gain dc amplifier designed to be stable over a wide range of feedback control. The gain of this amplifier is equal to the ratio of its feedback resistance to its input resistance as shown in Figure 2-25. If a diode is connected as the feedback resistance, as shown in Figure 2-26, the gain rises as the diode impedance goes up at low inputs and compensates for the low conduction at low signal levels. Figure 2-27 shows how the curvature in the passive silicon detector is

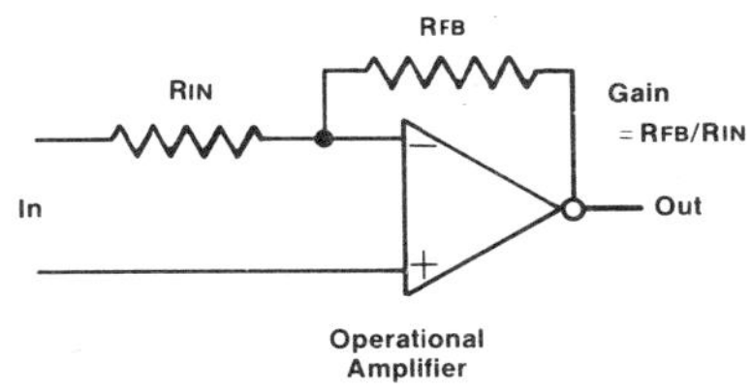

Figure 2-25—Gain of an Operational Amplifier

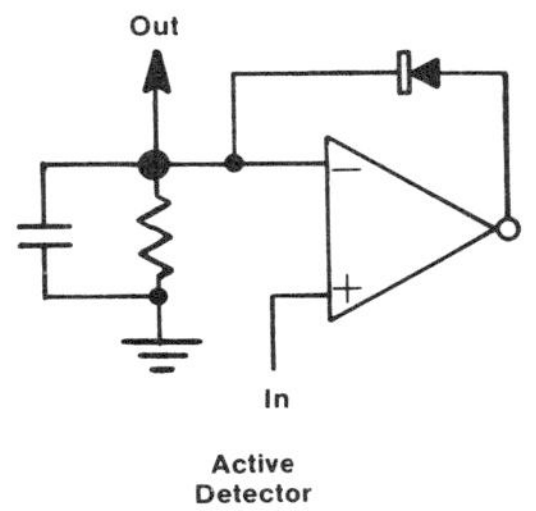

Figure 2-26—Diode Connected as Feedback Resistance in Op Amp

linearized in the active detector. One of the improvements in the art, made easily obtainable in the integrated circuit technology, is this active detector. It is used in radio receiver ICs with only a slight increase in circuit complexity, but a big improvement in performance.

Experiment: Set up a simple signal rectifying circuit as shown in Figure 2-17. Record and plot the output versus the input voltage. First, try various rectifier diodes without the operational amplifier. Suggested rectifiers include galena detector, germanium rectifier, silicon rectifier, carborundum detector, LED diode and silicon solar cell. Try various values of R and C. What is the effect of the relatively higher reverse leakage of germanium rectifiers and solar cells? Rectification efficiency is a function of the ratio of back resistance to forward resistance. In silicon it is high, accounting for generally superior performance of silicon rectifiers. How does the old galena crystal detector compare with a modern silicon rectifier?

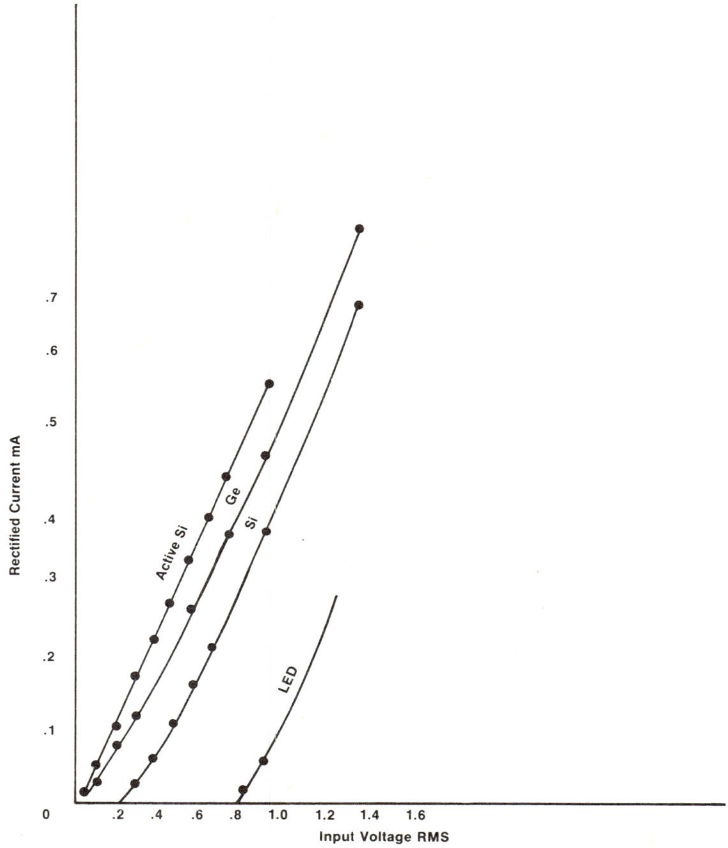

Figure 2-27—Linearization in Active Detector

Next, add the operational amplifier as shown in Figure 2-28 and repeat the experiments. Plot the results in comparison with the passive rectifiers as in Figure 2-29. Compare the frequency response of the active and passive rectifiers. Try them at radio frequencies. How do they compare? Remember that the active detector will be affected by the frequency response of the operational amplifier which is, unfortunately, basically a low-frequency device.

Figure 2-30 shows an active detector connected to demodulate a modulated signal generator.

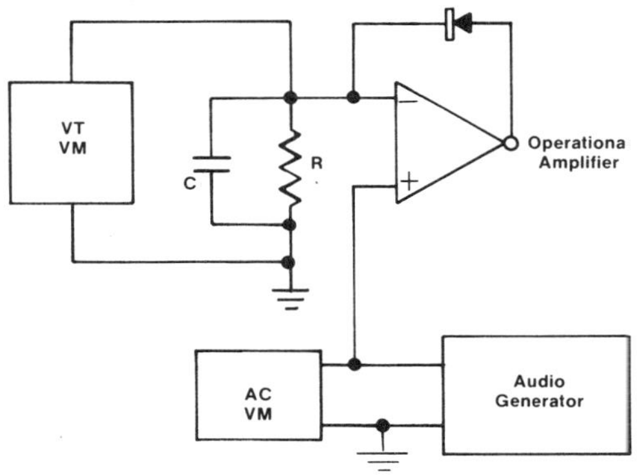

Figure 2-28—Signal Rectifying Circuit and Op Amp

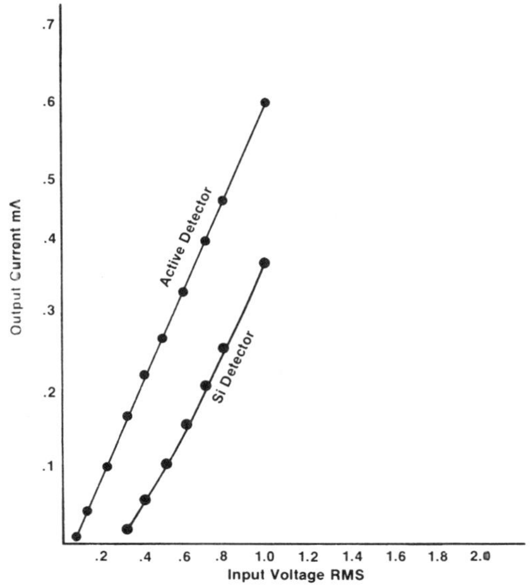

Figure 2-29—Active Detector Versus Si Detector

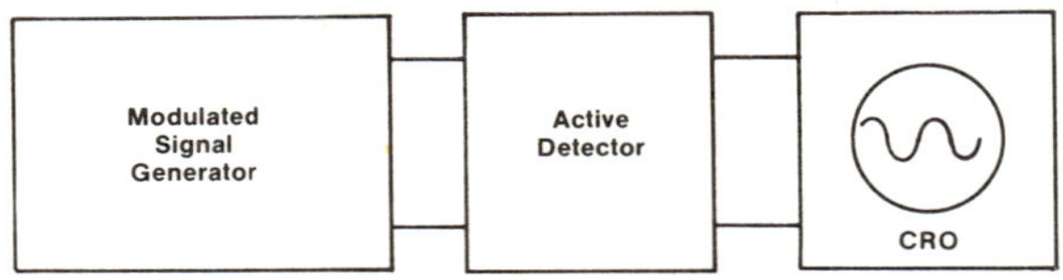

Figure 2-30—Active Detector Demodulating a Modulated Signal

Experiment: Set up the apparatus as shown in Figure 2-30. Observe the demodulated waveform as the percentage of modulation and the modulating frequency are increased. The advantage gained by the use of an active detector will depend on the carrier frequency and the frequency response of the operational amplifier used in the active detector circuit.

ZENER DIODES ARE A SPECIAL FORM OF DIODE

Diodes conduct in the forward direction. In other words, this happens when the anode is made positive with respect to the cathode, when the applied voltage is greater than a small threshold voltage (0.2 volt for germanium and 0.6 volt for silicon). In the reverse direction, there is very little conduction until it reaches a large reverse voltage where the diode breaks down and may become destroyed. However, some diodes start to conduct far below the breakdown voltage. This conduction can be controlled as to its sharpness and the voltage at which it occurrs. This latter is the basis for the zener diode, since this sudden conduction takes place at what is called the zener voltage. The zener diode, due to the fact that the zener conduction takes place at a very specific voltage in a given device is widely used as a voltage regulator. Figure 2-31 shows the complete conduction curve of a zener diode symbolized as shown.

The zener diode is often used as a shunt regulator, using the simple basic circuit shown in Figure 2-32. This same circuit can be considered a reference voltage source, the reference voltage being the zener voltage. Many more complex circuits are possible to improve the accuracy of a reference voltage. For example, since the zener voltage varies slightly with current, a more accurate reference is made possible by supplying the zener diode with a constant current.

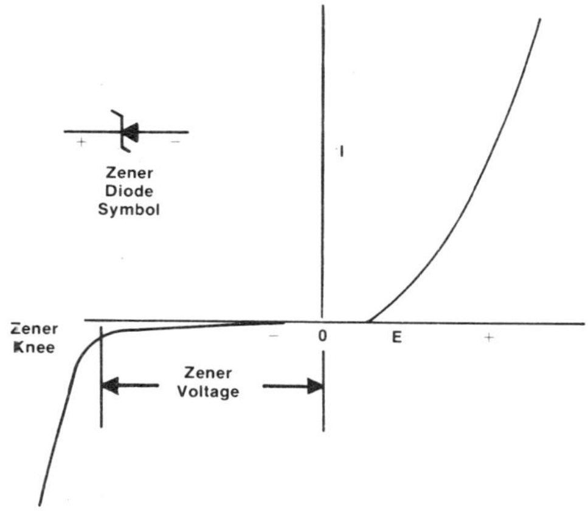

Figure 2-31—Complete Conduction Curve of a Zener Diode

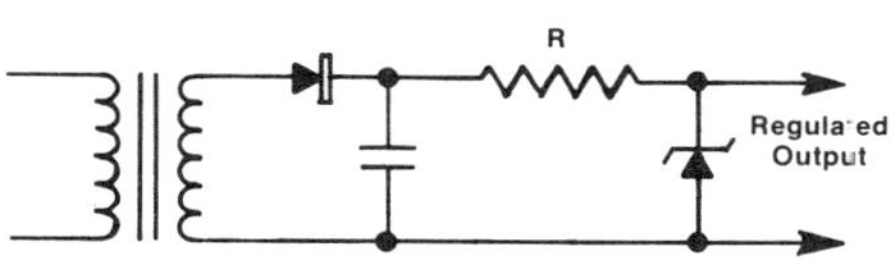

Figure 2-32—Zener Diode as Shunt Regulator

Zener voltages are attainable over a wide range, say, from about 3 volts to 100 or so. Many zeners are used having a zener voltage of 6.2 volts, since at this voltage the temperature coefficient is at a minimum.

PROJECT: ac/rf Ammeter

A multi-range ac/rf ammeter has a number of important uses, although such a meter is not readily available at a low price. The ammeter of this project is capable of measuring ac and rf currents from a fraction of a milliampere to 5 amperes. Thus, for example, the audio power from a power amplifier driving an 8 ohm load can be measured up to 200 watts ($I^2R = 5^2 \times 8 = 200$). The power output of an rf transmitter loaded with 50 ohms can be measured up to 1250 watts ($5^2 \times 50 = 1250$). Power to 120 volt line operated equipment can be measured up to 600 watts ($120 \times 5 = 600$).

The ac/rf ammeter consists of a series of low-resistance shunts which are selected by linking two binding posts on the front panel; a germanium rectifier bridge and a 50 microampere full scale meter. Ac or rf current to be measured is passed through the appropriate shunt and the resulting voltage drop is rectified and indicated on the dc microammeter. While a tap switch would be more convenient for changing ranges, no simple and inexpensive switch can safely carry the 5 ampere maximum current.

Figure 2-33 shows the complete schematic circuit diagram of the ammeter. An LED is shunted across the bridge and an SPST switch is connected in series with the meter. This provides a safety feature, since the meter switch is placed in its off position until the proper shunt is in place and current to be measured is flowing. If the LED lights, it shows that too much current is flowing in the selected shunt to be safely indicated by the meter, or that the binding post

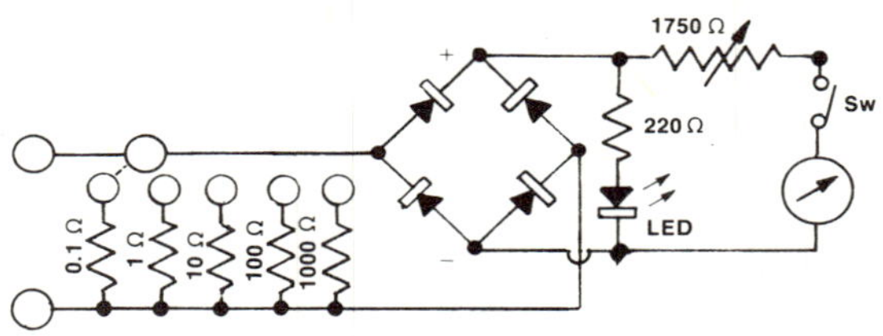

Figure 2-33—Circuit Diagram of Ammeter

link is not making good contact. Only if the LED does not glow should the meter switch be closed. The variable resister is used to calibrate full-scale sensitivity.

Figure 2-34 is a photograph of the completed meter. It shows a link connecting the center binding post to the one marked 5 (amperes). This places the 5 ampere shunt across the rectifier bridge and in series with the input and output binding posts.

Figure 2-35 is a photograph of the inside assembly and wiring of the ammeter. Wiring is simple and as direct as possible to cut down on series impedance. There is no need for a circuit board.

As in the case of all ac ammeters designed to provide the lowest possible series impedance, the rectifier bridge input voltage is quite low (0.5 volt full scale) and the resulting meter indication is nonlinear. It thus becomes necessary to provide a calibration or scale other than the linear scale of the dc meter. One can do either or both.

Figure 2-34—AC/RF Ammeter Complete

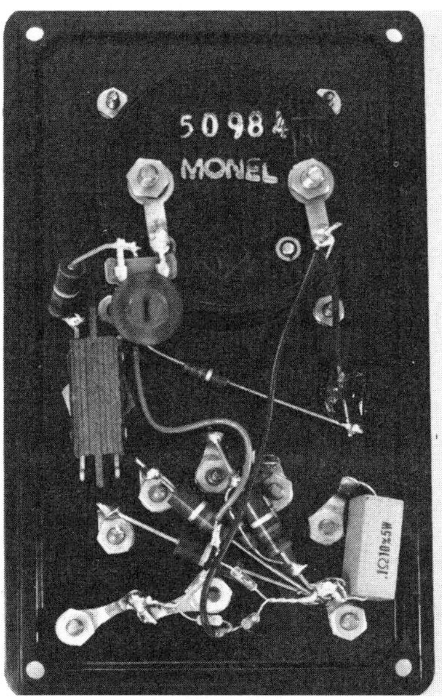

Figure 2-35—AC/RF Ammeter Inside

I Amps	Meter
0.5	1.0
1.0	2.0
1.5	5.0
2.0	9.0
2.5	15.5
3.0	22
3.5	29
4.0	37
4.5	45
5.0	52

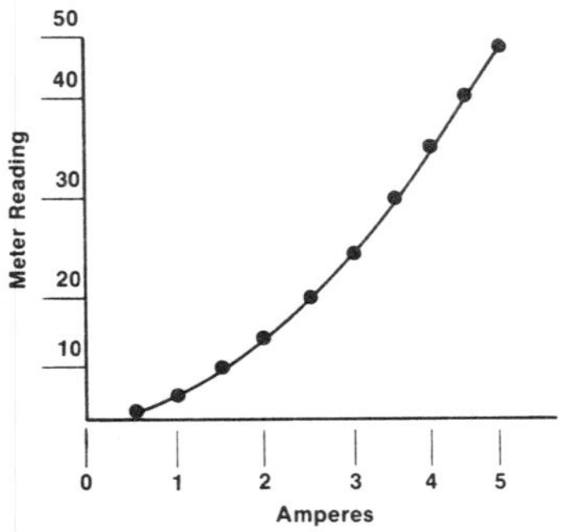

Figure 2-36—Ammeter Calibration Data and Plotted Results

Figure 2-36 is the calibration data and plotted results, which can be used rather than making the special scale.

ac/rf Ammeter Parts List

Description	Radio Shack Part No.
50 microampere dc meter	22-051
8 binding posts	274-661
SPST switch	275-324
0.1 ohm resistor	271-128
1.0 ohm resistor	271-073
10 ohm resistor	271-001
100 ohm resistor	271-012
1000 ohm resistor	271-023
5K potentiometer	271-1714
220 ohm resistor	271-015
Phenolic box and cover	Calectro H4-726

3

Projects and Experiments with Transistors

The step from the PN diode to the PNP transistor of today seems simple and obvious. However, it took many years and a great deal of experimentation to come up with the modern transistor. Diodes in the form of crystal detectors were known and used in the early 1900s. It was realized that if the tiny signals of wireless transmissions could be amplified, greater distances could be spanned with less power. Experiments were made on the crystal detector which came very close to the transistor. However, the vacuum tube came along providing the answer to the need for amplification and set the transistor back 25 years.

TWO DIODE JUNCTIONS MAKE A TRANSISTOR

When a PN diode is back-biased, little or no current flows. A potential barrier exists at the junction. The electrons in the N material are repelled by the reverse bias. Figure 3-1 is a greatly magnified picture showing how the reverse bias (−) on the point contact repells the electrons in the N material, preventing current flow. However, if a second point contact is placed close to the first point contact and is biased forward, current flowing to it will cause current to flow across the potential barrier to the first point as suggested by the arrows. This is basic transistor action. Specifically, it is a point contact PNP transistor. Symbolically, Figure 3-2 shows a PNP dual junction at A; drawn more like a transistor at B; and finally, the conventional transistor symbol at C. The arrow indicates the emitter, the rectangle the base and the angled line, the collector.

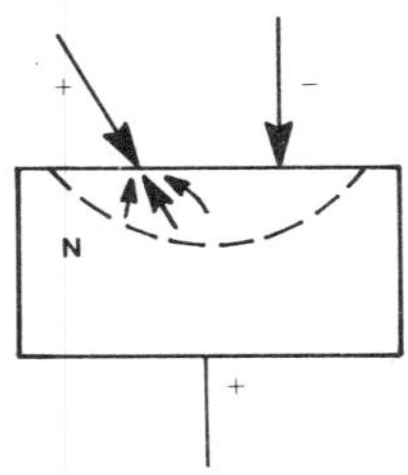

Figure 3-1—Reversed-Biased PN Diode

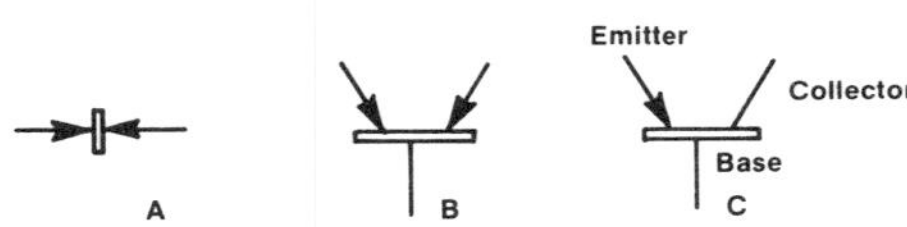

Figure 3-2—PNP Dual Junction, PNP Like a Transistor, Transistor Symbol

THREE BASIC TRANSISTOR CIRCUITS

A transistor acts as a valve allowing current to flow in the circuit of one pair of electrodes under the control of the third electrode. It is most often used as an amplifier, in circuits where a small current controls a substantially larger current.

The three basic transistor circuits are: common base, common collector and common emitter. One of the first transistor circuits was the common base as shown in Figure 3-3. This was because this seemed the most natural way to connect a point contact transistor, the first available type.

Experiments with Common Emitter Transistor Circuits

The most widely used transistor amplifier circuit is the common emitter circuit as shown in Figure 3-4. First, with $R_2 = 1000$ ohms and R_1 variable from about 50K to 1M take the static characteristic, observing I_C as I_B is varied from say 10 microamperes to 100 microamperes. Also, observe E_B. Plot the results as shown in Figure 3-5. Now, if an ac input signal is applied, the collector current will vary around the dc bias point, say the **mid-point** corresponding to a dc bias current of 33.5 microamperes. Choosing this mid-point permits the maximum positive and negative excursions of collector current in which the collector voltage swings between zero and the bias voltage, Vcc.

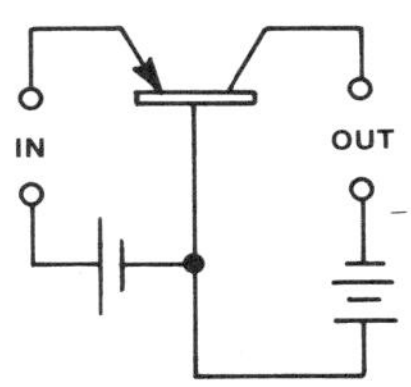

Figure 3-3—Common Base Transistor Circuit

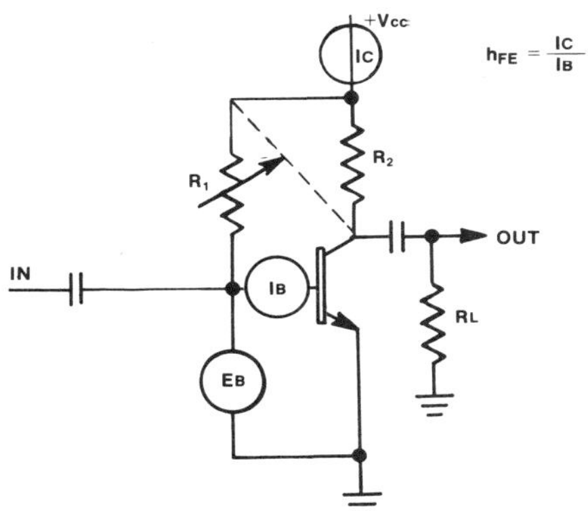

Figure 3-4—Common Emitter Circuit

Stabilizing the Common Emitter Amplifier

The plot of Figure 3-5 indicates that the mid-point at 4.75 volts at the collector corresponds with a base current of 33.5 microamperes. However, this is true only for this particular transistor at a room temperature of 25° C. What about other transistors and other temperatures?

The particular transistor used to generate Figure 3-5 had a current gain of 143 (h_{FE}) at a collector current of 5 ma (5 ma/35 microamperes). When this transistor was heated to 100°C, its collector current increased to 7.5 ma or an apparent current gain of 214 (see arrow A on Figure 3-5). This means that whereas the collector current can swing approximately 5 ma above and below the mid-point, it can only swing a little less than 2.5 ma above point A without causing distortion due to running out of collector voltage.

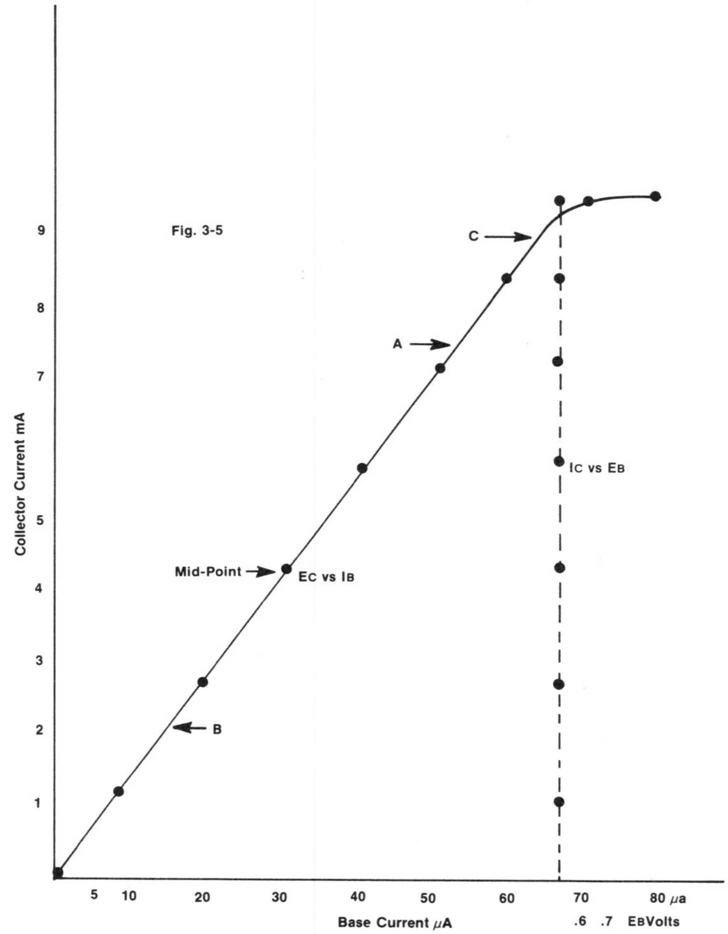

Figure 3-5—Collector Versus Base Current in Common Emitter Circuit

Another cause of operation away from the mid-point is the differing current gain characteristics between different transistors. Transistors are often specified for "typical" current gain. However, transistors bearing the same type number can exhibit widely different current gains. In Figure 3-5 one sample transistor operated at mid-point while another transistor having the same designation showed a much higher current gain, and with the same base current had a collector current of 9 ma (point C). Still another had a collector current of 2.1 ma (point B). It can be easily seen how the output voltage swing of these latter two transistors will be severely limited.

Experiments are indicated here to determine the effect of differing current gains in transistors. There are two important ways in which the problem of divergent current gains can be countered. One is to use a lower value of collector resistor and the other is to return the base bias resistor to the collector instead of to the bias voltage point, as shown by the dotted line in Figure 3-4.

Experiments with Common Collector Transistor Circuits

The common collector circuit is a simple, unity gain, impedance transformer. The basic circuit is shown in Figure 3-6. The maximum swing of output voltage is possible when the emitter current times the emitter resistor (R_2) is equal to one-half the collector supply voltage (V_{CC}). The gain of the stage is just slightly less than unity.

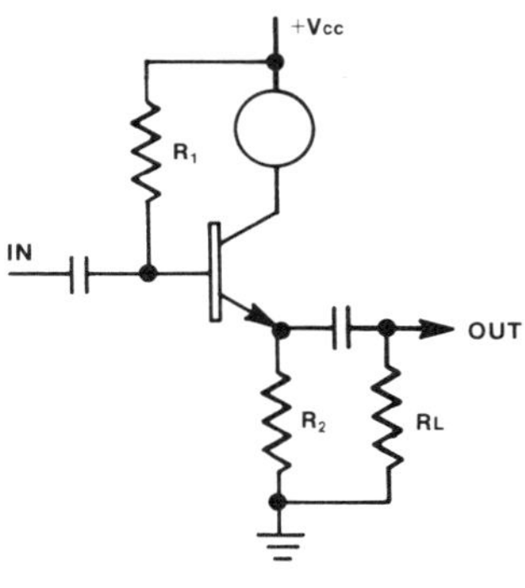

Figure 3-6—Common Collector Circuit

Measurement of input impedance is easily carried out by merely placing a variable resistor in series with the input and varying it until the output is halved. Of course, this input resistance is actually the input resistance of the transistor in parallel with the base bias current resistor R_1.

Output impedance is not so readily measured. From a practical standpoint, if R_1 is varied until the output voltage starts to distort, the minimum tolerable load resistance will be indicated.

This circuit will be seen to be highly degenerative, and hence is not sensitive to differing transistor gain characteristics. It makes a useful impedance transformer for coupling a high impedance source to a low impedance sink.

Common Collector as a Current Amplifier

The common collector circuit is widely used as a current amplifier. This is compatible with the low output impedance characteristic of this circuit.

In Figure 3-7, a common collector transistor is connected to drive a loudspeaker. When two transistors are used in complimentary symmetry or quasi-complimentary symmetry circuits, the typical power output circuit is created as described in Chapter 10.

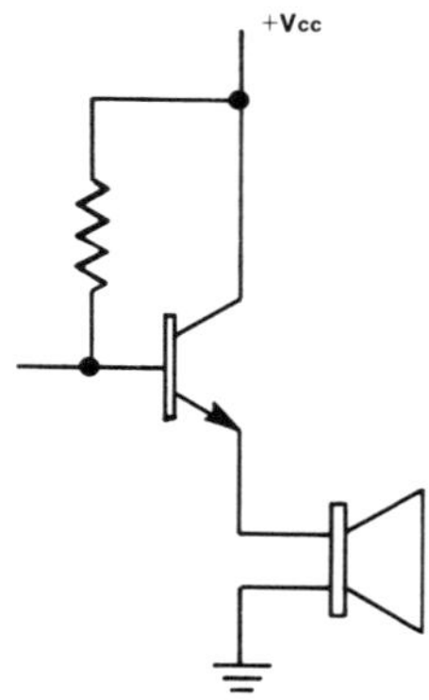

Figure 3-7—Common Collector Circuit Hooked to Loudspeaker

Experiments with Common Base Transistor Circuits

The third variation in transistor circuits is the common base circuit as shown in Figure 3-8. The base bias is applied between base and ground, and the emitter (and collector) current is $(E_B - E_{BE})/R_1$. For example, to bias this circuit for a collector current

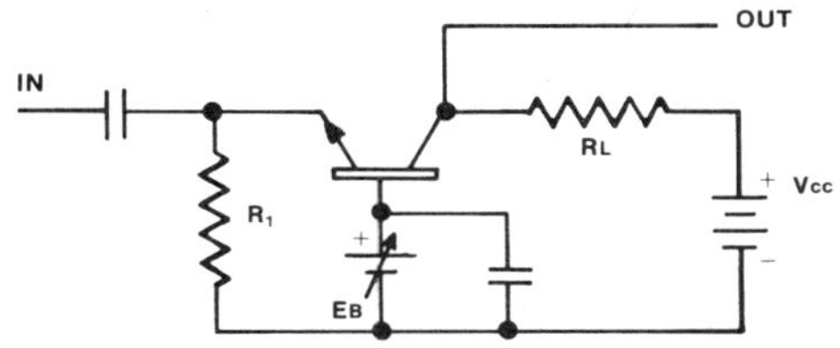

Figure 3-8—Common Base Circuit

of 5 ma, a base bias of 1.5 volts, less a base-to-emitter drop of say 0.5 volts, will supply 5 ma through 200 ohms (R_1). The ac input impedance of this circuit is very low (on the order of 10 ohms); the output impedance is high and the voltage gain is high.

Experiment by measuring ac voltage gain and input and output impedance.

The common base circuit has often been used as a radio frequency amplifier due to its inherent stabiltiy. Since the base is operated at ground potential (for rf) it acts as a shield between emitter and collector, generally reducing feedback from output to input.

A two-stage tuned rf amplifier circuit is shown in Figure 3-9. This circuit, when fed to a detector and audio amplifier, has plenty

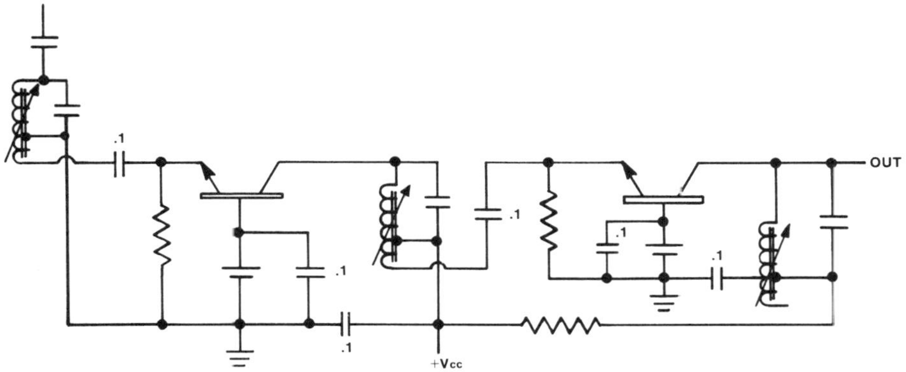

Figure 3-9—Two-Stage Tuned rf Amplifier Circuit

of gain and selectivity for a high fidelity local radio receiver. The three coils should be mounted in separate shielded compartments with the transistor sockets mounted on the partitions between the first and second coil compartments, and between the second and third coil compartments. The turns ratio between the tuned portion of the coils and the untuned portion is 10—1. The coupling and bypass capacitors are all 0.1 microfarad ceramic or tantalum.

PROJECT: Build a Transistor Checker

One of the most useful pieces of equipment one can have when engaged in experimenting and project building is a transistor checker. A simple and easily constructed transistor checker will measure the current gain (h_{FE}) of a transistor and whether it is open or shorted. Excessive leakage current between collector and base can also be indicated. The ratio of the collector current to the base current is the dc current gain or h_{FE}, the most important single parameter of any transistor.

The circuit of the transistor checker to be described is shown in Figure 3-10. It provides for checking either NPN or PNP transistors by merely throwing a double pole-double throw toggle switch. A meter measures collector current, I_C, that is, the collector-to-emitter leakage with the base open. A germanium transistor will show a few microamperes of leakage but a silicon transistor leakage should be too small to read on the 1 milliampere meter. Shorted or excessively leaking transistors are spotted in this test. Keeping push-button A depressed, push-button B is pressed supplying 10 microamperes (6-0.5/550K) to the base. The meter will read h_{FE} with a full scale reading of 100 (1000/10 = 100). General purpose low-power transistors should read around 30 to 100. Power transistors (connected by means of the tip jacks) will read on the low side due to the fact that h_{FE} depends on collector current, and power transistors designed for high collector current show low h_{FE} at the 10 microampere base current. High gain transistors may well have a current gain of over 100 (off scale), so a 5X multiplier is provided. When this push-button is pressed (button C) the scale is multiplied by 5 so that the meter reads h_{FE} of 500 full scale. Open transistors show no indication on the meter. Shorted transistors show excessive I_{CO}. The meter is protected against shorts by an LED (this form of meter protection is described in Chapter 2).

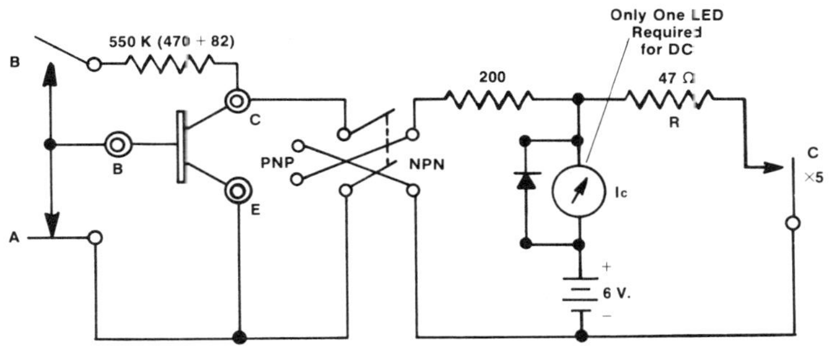

Figure 3-10—Transistor Checker Circuit

TRANSISTOR CHECKER PARTS LIST

(Note: Some items are packaged in quantities greater than required for this project).

Description	Radio Shack Part No.
6 Transistor Sockets	276-548
5 Push Button Normally Open	275-1547
5 Push Button Normally Closed	275-1548
1 DPDT Toggle Switch	275-614
4 AA Battery Holder	270-391
1 ma Meter	22-052
3 Phone Tip Jacks	
4 AA Batteries	
Phenolic Box 6 × 3½ × 1⅞	Calectro H4-726

Radio Shack Part No. 270-627 is a bakelite box with an aluminum panel. I prefer the Calectro box which has a black phenolic panel. The dimensions in Figure 3-11 are based on the Calectro panel.

Figure 3-11 shows the drilling pattern and dimensions for the

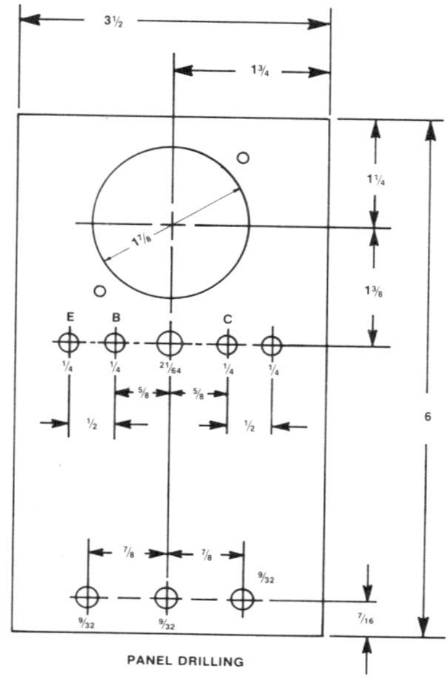

Figure 3-11—Drilling Pattern for Front Panel

front panel. These dimensions are to be adjusted in accordance with the outside dimensions of the actual panel used.

Figure 3-12 shows the assembly and wiring viewed from the back. It is good practice in wiring projects to use colored wire in which each color is assigned to a particular circuit. This helps in checking to see that all circuits are correctly wired, and again in tracing circuits in case of trouble. The following table contains suggestions for color coding of projects:

Color	Circuit
Red	Positive battery
Black	Ground
Blue	Collector
Orange	Emitter
Green	Base
White	Negative battery

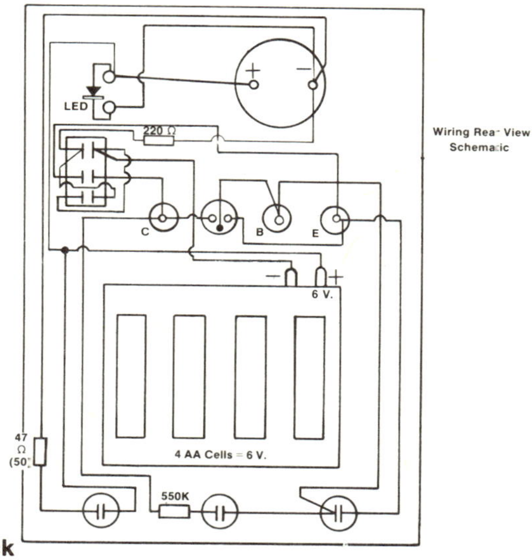

All leads to
+ Meter = red
− Meter = black
to C = blue
to B = green
to E = orange

Figure 3-12—Transistor Checker Panel Wiring

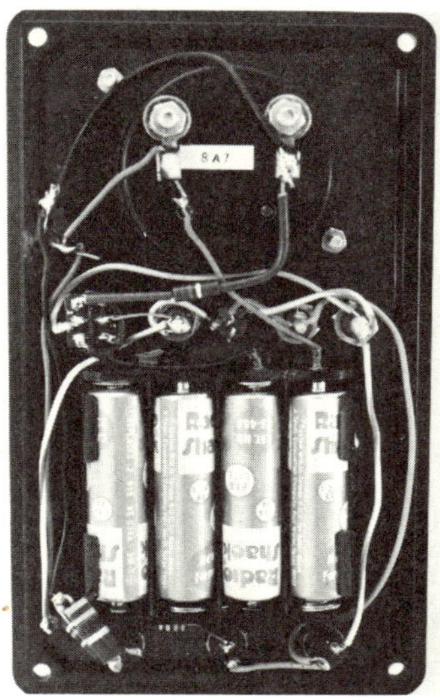

Figure 3-13 is a photograph of the completed transistor checker. Attractive panel markings can be provided by printing with a Dymo labeler or the equivalent using black tape (or the color of your choice).

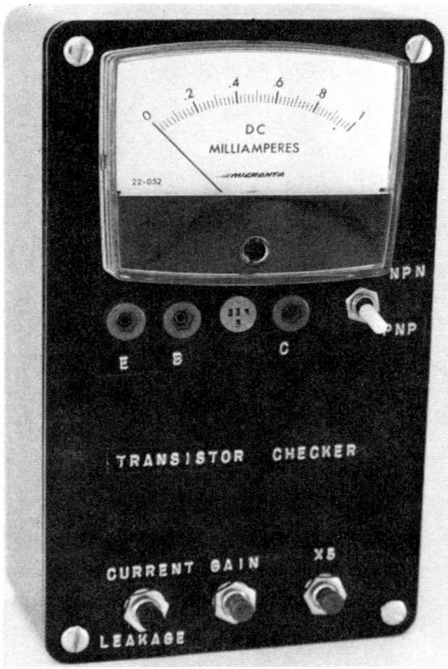

Figure 3-13—Transistor Checker Completed

CASCODE AMPLIFIER EXPERIMENTS

The term cascode was first applied to a vacuum tube amplifier. As shown in Figure 3-14 it consisted of a common emitter triode circuit followed by a common grid amplifier. The original circuit was developed in an attempt to reduce the noise of a low-level amplifier while simulating a pentode characteristic. Pentodes were developed for rf amplifiers in order to do away with the complications of neutralization, but they generated considerably more noise than triodes. The cascode amplifier showed low noise and did not require neutralization due to the grounded grid-isolating effect between the cathode and plate of the output stage.

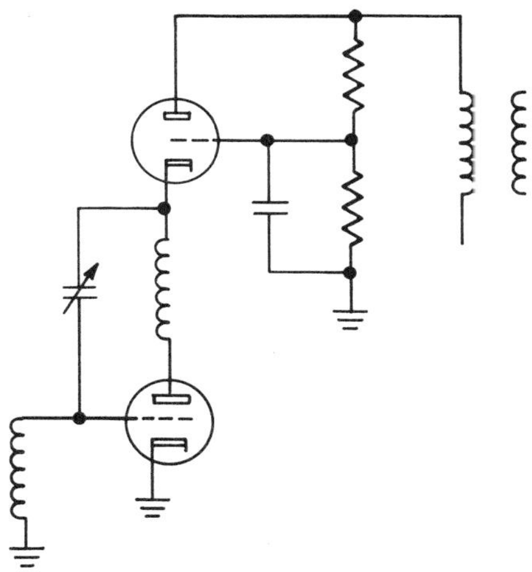

Figure 3-14—Circuit for a Cascode Amplifier

The cascode circuit transistorized is shown in Figure 3-15. The first transistor acts as a current amplifier, while the second acts as a unity gain repeater. Thus, the overall gain of the two transistors is $h_{FE} \times R_L$. If the input is measured in volts, the gain is $G_M \times R_L$. The second stage provides input to output isolation as in the rf amplifier of Figure 3-9 while the added first stage increases the input resistance from a few ohms to about 1000 ohms. One additional feature of the cascode circuit is that the first transistor is operated at a constant collector-to-emitter voltage. This latter has the effect of reducing distortion when the cascode circuit is used as an audio amplifier.

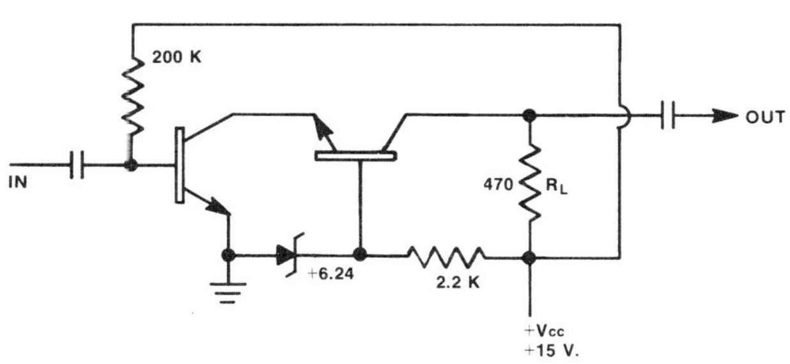

Figure 3-15—Transistorized Cascode Circuit

Experiments: Build the circuit of Figure 3-16 on a small perf board. With the switch in position A, measure gain input resistance and frequency response of the cascode circuit. Repeat the measurement with the switch in position B, as a common emitter circuit. Compare distortion of the two circuits. Substitute the cascode circuit for the common base amplifier in the rf amplifier of Figure 3-9. The gain and selectivity should be increased due to the increase in input resistance.

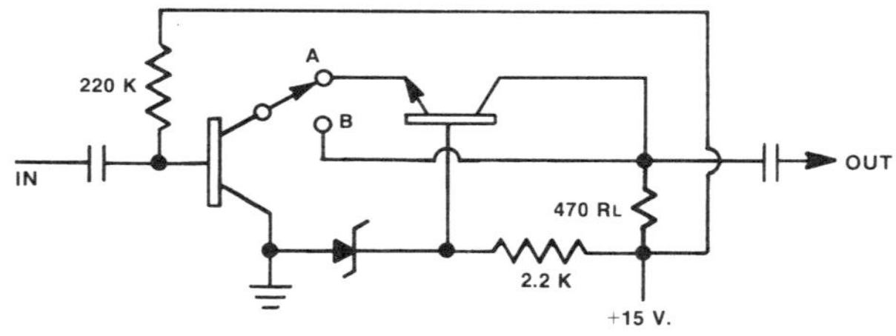

Figure 3-16—Cascode Circuit for Experimentation

The constant voltage across the collector to emitter of the input transistor is provided by the emitter-follower action of the second transistor (the emitter voltage follows the fixed base voltage determined by the zener diode). This mode of operation can be used in an emitter-follower circuit as shown in Figure 3-17. Compare the distortion in the load of this circuit with a simple emitter follower by omitting the common base transistor.

Using power transistors and a Darlington input driver provides a low distortion power amplifier. This circuit can also be developed

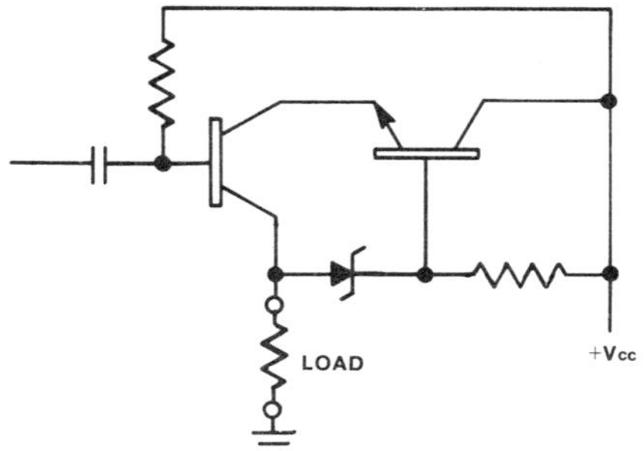

Figure 3-17—Emitter-Follower Circuit

into a low distortion complementary symmetry power amplifier with an extended high frequency response. Low transient intermodulation distortion requires this latter.

DARLINGTON PAIR EXPERIMENTS

One of the most useful of all transistor circuits is the Darlington pair. In the Darlington circuit, as shown in Figure 3-18, a first transistor drives a second transistor directly from emitter to base. The input transistor provides current gain and, therefore, a substantially increased input resistance.

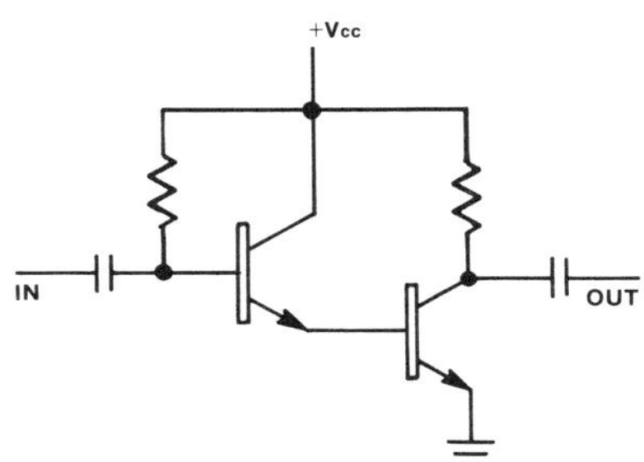

Figure 3-18—Darlington Pair Transistor Circuit

The circuit of Figure 3-19 is a Darlington pair providing high current gain and unity voltage gain. It is a very useful power output circuit for driving low impedance loads such as loudspeakers.

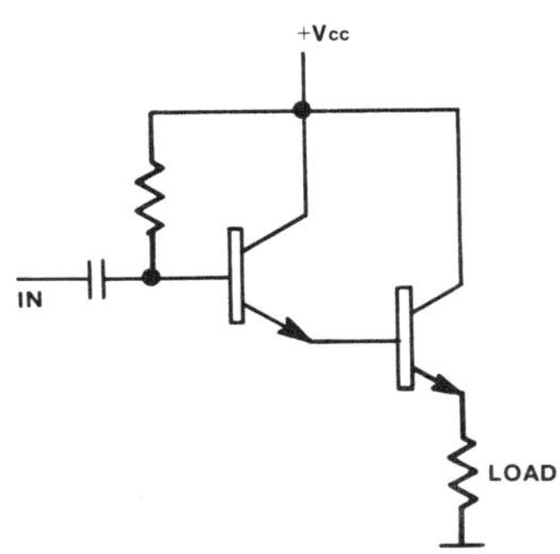

Figure 3-19—Darlington Pair with High Current Gain

Combining it with a second similar circuit of opposite polarity creates a complementary symmetry power output circuit which is in wide use.

The circuit of Figure 3-20 is similar to that of Figure 3-18, but with the addition of ac feedback. This feedback improves waveform and stability and makes the circuit less critical in the selection of transistors.

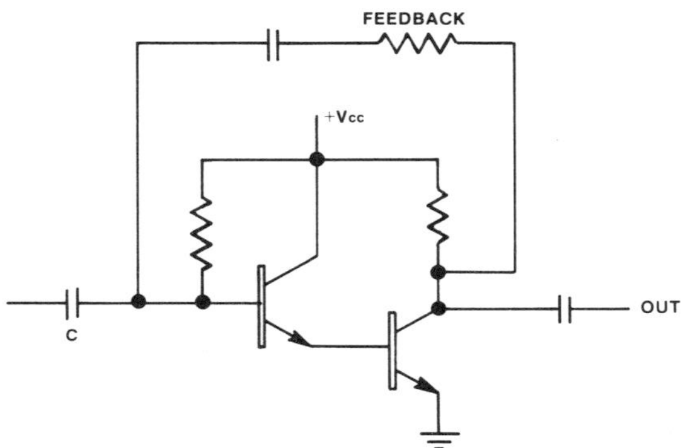

Figure 3-20—Darlington Pair with ac Feedback

Experiments: Set up the three circuits of Figure 3-18 through 3-20 and measure their characteristics. Measure gain and frequency response versus the collector resistors in Figures 3-18, 3-19 and 3-20. Take care in each case to select the collector to base bias resistor of the input stage so that the collector of the output stage sits at a voltage substantially equal to one-half the positive bias voltage. Check the input impedance in each case to see how it changes with the different circuits.

One way to easily measure the input resistance of these transistor circuits is to lower the test frequency until the output falls to 0.7 of maximum. At this point, the impedance of the input coupling capacitor is equal to the input resistance ($1/C\omega = R$).

EXPERIMENTING WITH SOLID STATE SWITCHES

The transistor, operated as a solid state switch, made digital integrated circuits possible. In addition to the wide use in digital logic circuits, the transistor is also widely used as a circuit-switching device. In both applications, small size and fast operation are the most important features. In addition to transistors, solid state switches include diodes in simple and complex configurations. When switches are combined, gates, flip-flops and other logic building blocks are formed. Most switches have two states, namely, ON or OFF although a special type called a three state switch has a third, or passive state.

The diode as a switch has some important, though limited, applications. The best example is as a rectifier in a power supply where it passes current in one direction and blocks it in the other as shown in Figure 3-21.

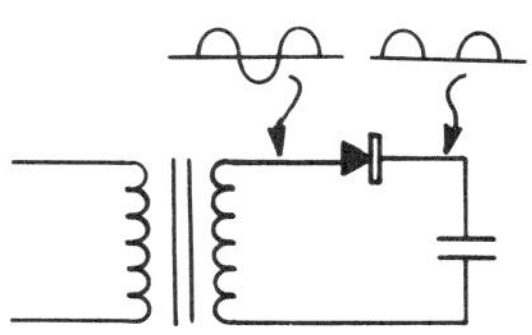

Figure 3-21—Diode as a Rectifier

The efficiency of a diode used as a switch is a function of the ratio of its forward resistance, R_F, to its reverse resistance, R_R, as illustrated in the typical diode I/V plots of Figure 3-22. Low forward resistance and high reverse resistance are the desirable features. **Experiment:** Measure and plot the current versus voltage, forward and reverse, of a germanium diode and a silicon zener diode using the set-up of Figure 3-23. The slopes (E/I) are the resistances as noted on the plots. However, these plots and computations assume a zero resistance current meter and must be corrected by subtracting the current meter resistance to obtain the actual diode resistances as shown on Figure 3-17. It should also be noted that in measuring such steep slopes, the ability to read small increments of voltage is also important.

Figure 3-24 shows how the forward resistance of diodes and the low resistance of zener diodes can be measured more accurately. The currrent meter is placed before the voltmeter so that its resistance is not included in the diode resistance. Voltage measured is bucked to enable one to use a low voltage scale,

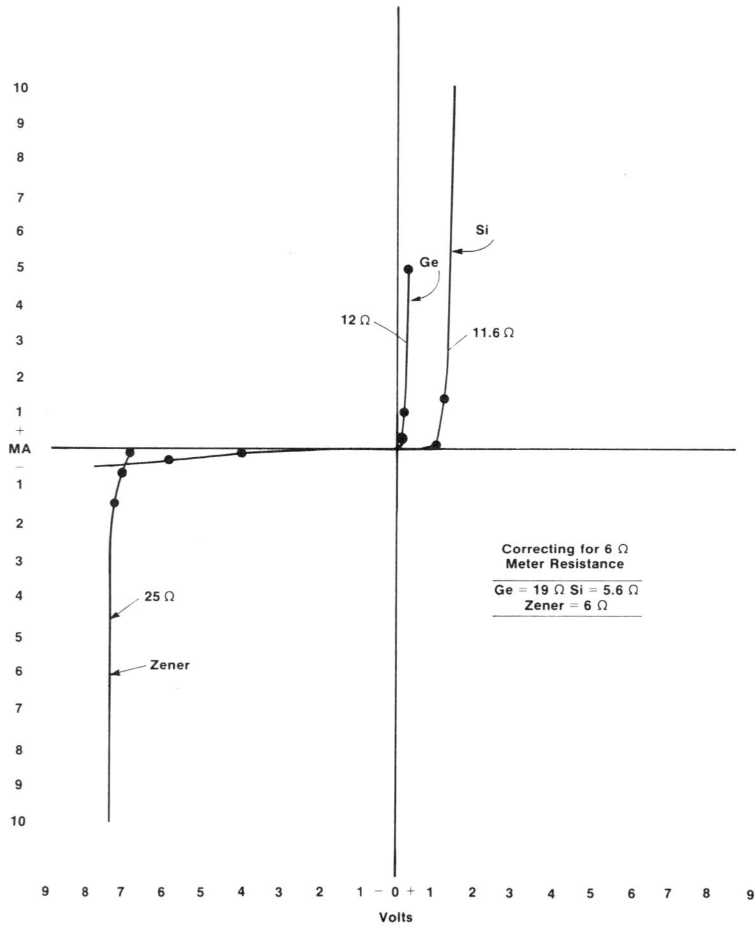

Figure 3-22—Diode Efficiency Curves

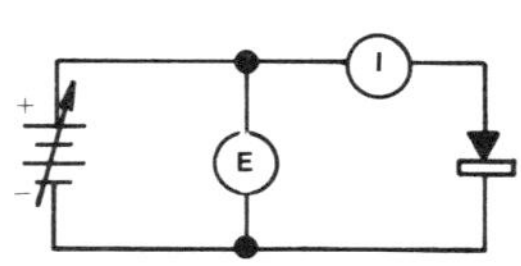

Figure 3-23—Ge Diode and Si Zener Diode Circuit

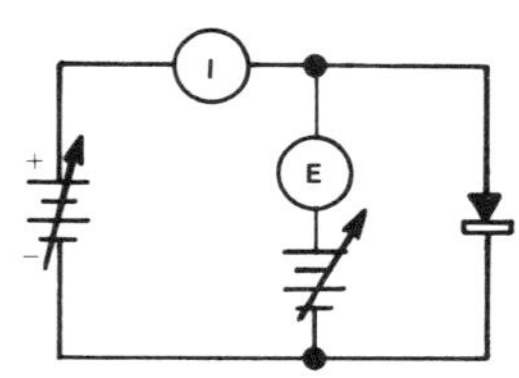

Figure 3-24—Improved Test Circuit

making the measurement of increments more accurate. If the resistance is measured by observing the voltage change for a change of diode current from, say, 5 to 10 milliamperes, a voltmeter drawing 50 microamperes full scale will introduce very little error. The two adjustable voltage sources should be two adjustable regulated power supplies.

Measuring the ON and OFF Resistances of Transistor Switches

The above experiments with diodes illustrate ways to measure diode conduction and zener impedances. The impedance of a transistor as a switch can be measured in much the same way. **Experiment:** Set up a transistor circuit as shown in Figure 3-25. As the base current is increased, observe and record the collector current and voltage. Plot the collector-to-emitter volts and equivalent resistance. Measure the impedance collector-to-emitter with the base grounded. Set up a circuit to drive the transistor base with a square wave and observe the collector to emitter voltage on a cathode ray oscilloscope as shown in Figure 3-26. This will demonstrate the limitations on switching speed.

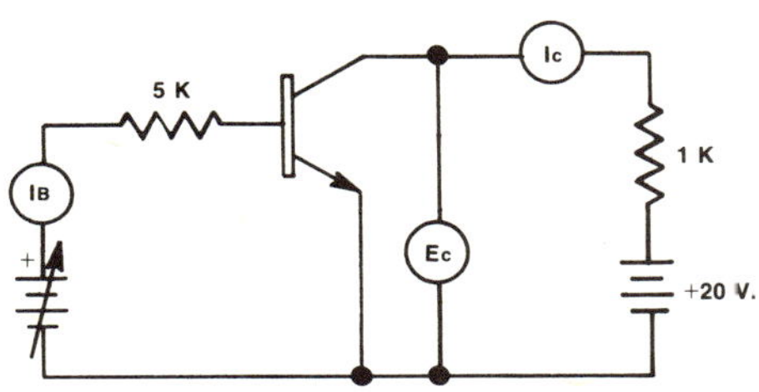

Figure 3-25—Circuit for Measuring ON/OFF Resistance of Switches

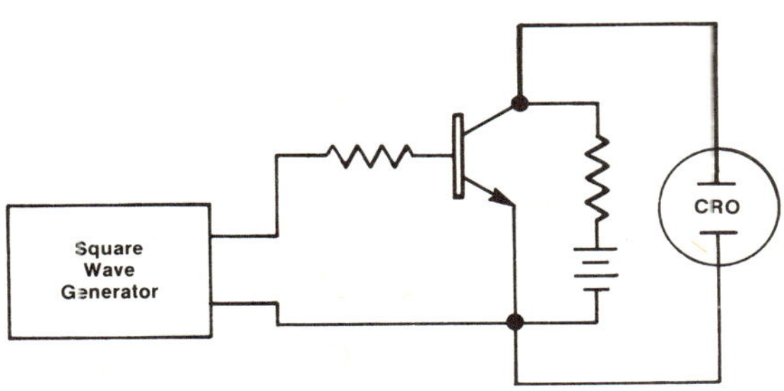

Figure 3-26—Driving Transistor with a Square-Wave Generator

The important characteristics of a switch are: (1) its impedance when OFF: (2) the voltage it will support when OFF: (3) its impedance when ON: (4) its voltage drop when ON: (5) the current it will support without overheating when ON; (6) its switching speed both going ON and going OFF.

It is important to observe that the dissipation in a switching transistor is relatively low in both ON and OFF states. This, for example, is responsible for the high efficiency of switching regulators. On the other hand, the transistor is called on to dissipate power when switching from one condition to the other and this dissipation increases with switching frequency.

4

Projects and Experiments with Solid State Power Supplies

POWER SUPPLY REGULATION AND RIPPLE

In Chapter 1, a simple adjustable power supply was presented as a project. The characteristics of this power supply determine its range of application. The following comments are made in reference to this power supply. The 120-volt 60-hertz line is transformed to 6.2 volts by the line transformer. Hertz means cycles or oscillations per second. The secondary voltage is rectified by a single diode rectifier, providing what is known as half-wave rectification. This means that the positive half-cycles from the transformer secondary are passed on to charge the filter capacitor in a positive direction while the negative cycles are blocked. The rectified direct current (dc) appears across this filter capacitor. Every 1/60th of a second the capacitor is charged to the peak voltage of the secondary or $6.2 \times 1.4 = 8.68$ volts. If a load resistor or device to be powered is connected across the capacitor, it will cause the capacitor to discharge between peaks of the charging voltage. This fluctuation of voltage caused by the load is called ripple. The magnitude of the ripple depends mainly on the capacity of the filter capacitor and the current drawn by the load. Figure 4-1 shows how the charging and discharging of the filter capacitor generates ripple voltage. the ripple can be measured by means of a peak-reading voltmeter and can be observed on a cathode ray oscilloscope.

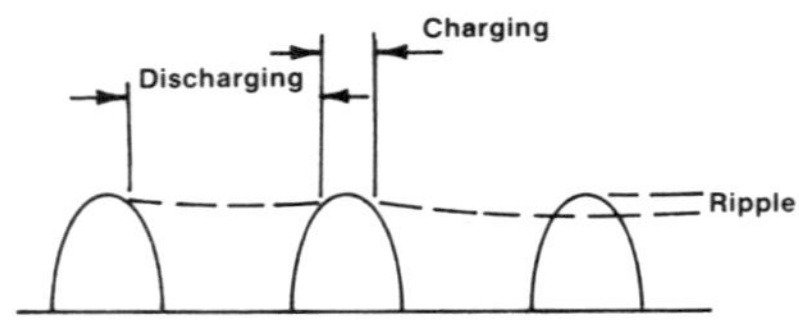

Figure 4-1—Capacitor Voltage in Half-Wave Rectifier Supply

Even more basic are the voltage and current characteristics of a power supply called its regulation. As the load resistance is reduced, the output current increases and the output voltage decreases. The ripple also increases. Figure 4-2 is a representative showing of the regulation and ripple characteristics of an unregulated power supply.

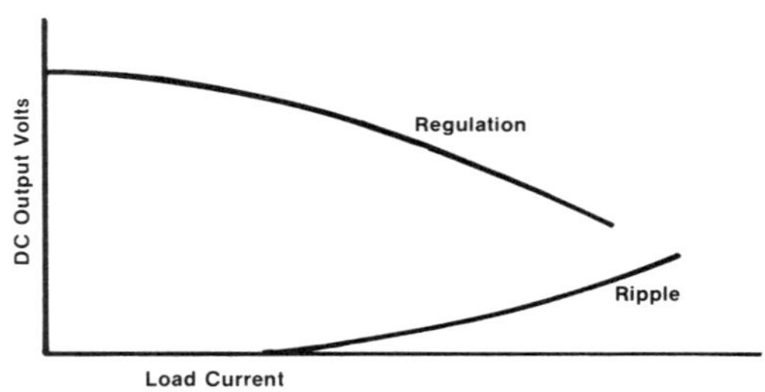

Figure 4-2—Regulation and Ripple Characteristics of Unregulated Power Supply

SEVERAL SIMPLE POWER SUPPLY CIRCUITS

The power supply of Chapter 1 is adjustable by virtue of the transistor connected in series with the output. This transistor is a convenient way of obtaining a variable resistor, adjustable over a wide range of resistance and, at the same time, capable of carrying substantial current. When this transistor is controlled by suitable reference and feedback means, it becomes the basis of a typical series-regulated power supply.

A line transformer having a single, untapped secondary can be used in two ways. First, with a single rectifier in a half-wave rectifier circuit as shown in Figure 4-3; and second, using a bridge rectifier in a full-wave rectifier circuit as shown in Figure 4-4. The full-wave circuit provides reduced ripple.

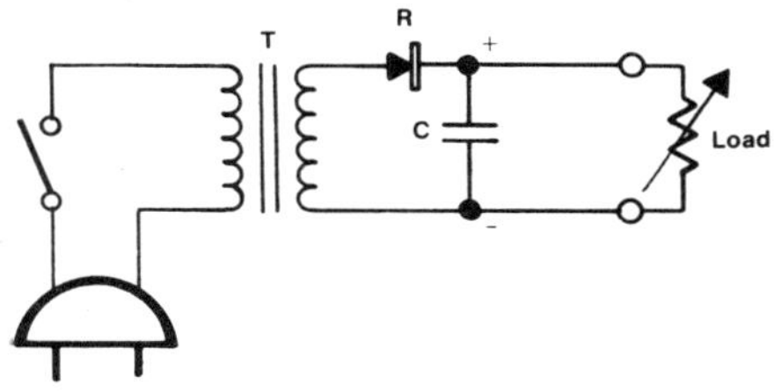

Figure 4-3—Single Rectifier in a Half-Wave Rectifier Circuit

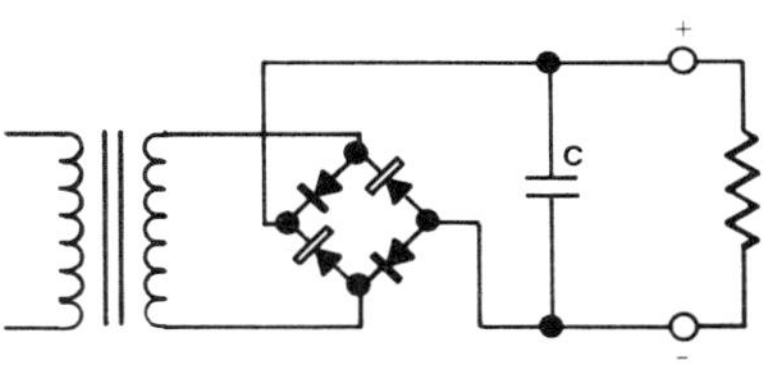

Figure 4-4—Bridge Rectifier in a Full-Wave Rectifier Circuit

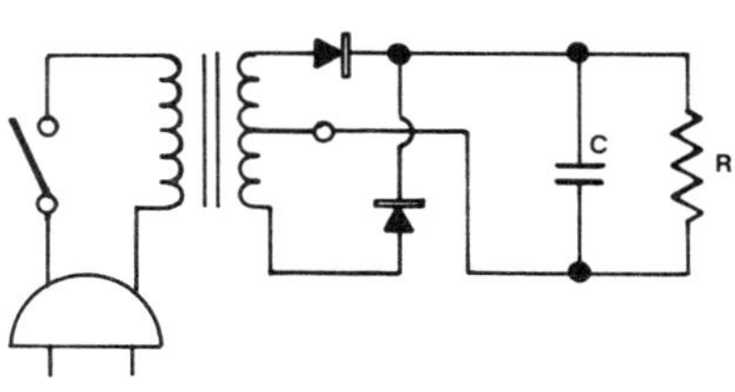

Figure 4-5—Full-Wave Rectified Power Supply

A more flexible transformer is one with a center-tapped secondary. With such a transformer the full-wave rectified power supply shown in Figure 4-5 is possible. The center-tapped transformer also makes possible a power supply having equal positive and negative outputs with reference to the center-tap as shown in the circuit of Figure 4-6. This circuit not only supplies positive and negative voltages, required by many solid state circuits, but it is also a full-wave rectifier circuit.

Experiments: Build the circuits set forth above and measure output voltage, regulation and ripple. The regulation is a function of the source resistance, the resistance of the transformer secondary and the rectifier forward conduction resistance. The ripple is mainly a function of filter capacitor capacitance and load resistance. Compare germanium with silicon rectifiers. Keep in mind that germanium rectifiers are much less heat tolerant than silicon. Heavy duty rectifiers require heat sinks to keep their temperatures at safe levels.

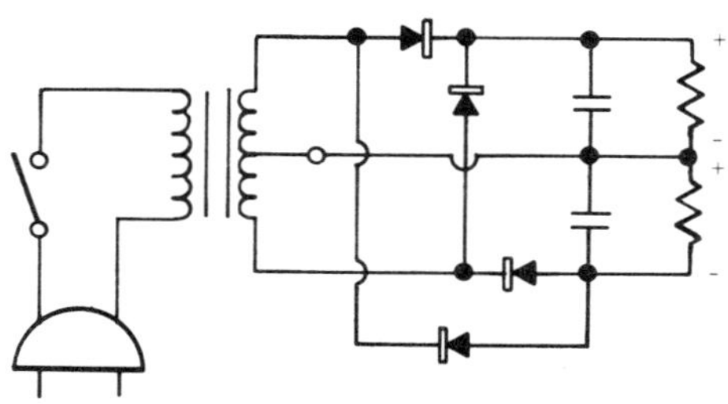

Figure 4-6—Power Supply with Equal Positive and Negative Outputs

Figure 4-7 is a circuit of the basic set-up for the power supply experiments. The transformer is a step-down transformer rated at 120 volts to 12.6 volts with the secondary center tapped (6.3 volts each half). The rectifiers and load resistors for the experiments are clipped into the circuit or may be wired in and soldered temporarily.

Experiments can be performed using the circuits shown above and taking the data of output voltage (converted into current) and ripple as illustrated in Figure 4-2.

On the more experimental side rather than the purely practical, other devices than the usual power rectifiers can be used to convert ac to dc. As will be explained more fully below, many solid state devices are basically PN diodes and, as such, can be used as rectifiers. Zener diodes, LED diodes and solar cell diodes are examples. This can be readily demonstrated with the set up described above.

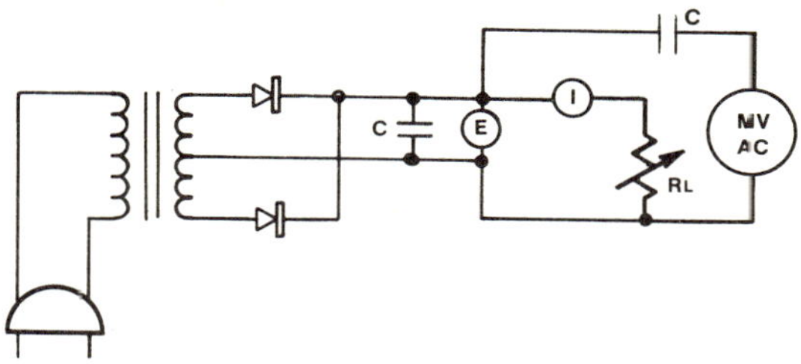

Figure 4-7—Basic Set-Up for Power Supply Experiments

The data shown above is for germanium rectifiers (Ge) type 1N91 or 1N92. It is interesting to substitute silicon (Si) rectifiers and to repeat the experiments. Then answer such questions as: Why are silicon rectifiers more often used when germanium are more efficient? Are there cases where germanium are definitely to be preferred? What are the limitations of each type?

VOLTAGE MULTIPLIER POWER SUPPLIES

There is a class of voltage supply circuits called voltage multipliers, which can be of use where higher voltages at limited currents are required. Regulation is not as good, but for low currents they can be useful. These circuits make interesting subjects for experimentation and will be shown without further comment. They all operate from a single source of ac voltage, and in the case of the set-up used above, the entire 12.6 volt secondary can be used as the input. Figure 4-8 is a half-wave doubler while Figure 4-9 is a full-wave doubler. It is interesting to compare the advantages and disadvantages of each. Figure 4-10 is a tripler while Figure 4-11 is a

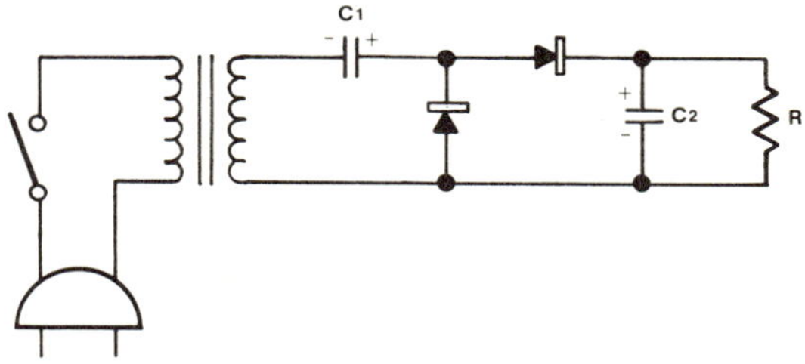

Figure 4-8—Half-Wave Doubler

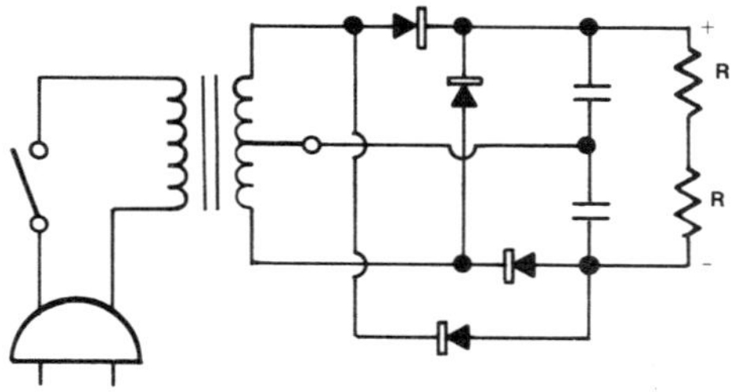

Figure 4-9—Full-Wave Doubler

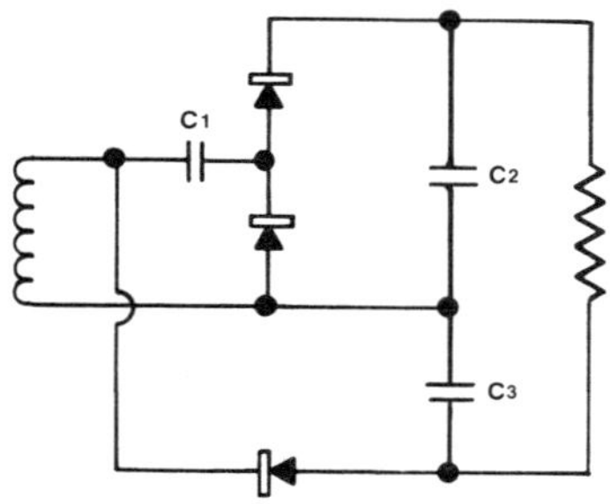

Figure 4-10—Tripler

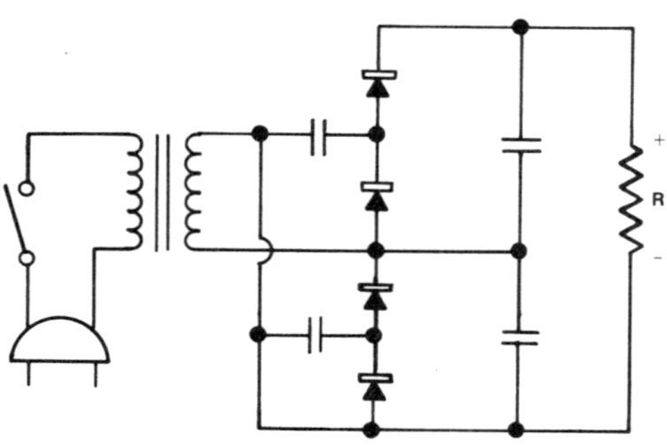

Figure 4-11—Quadrupler

quadrupler. It should be mentioned that not all capacitors in these circuits support the same voltages and this must be taken into consideration in building the circuits.

One of the most significant ratings of power supply rectifiers is their peak inverse voltage usually designated PIV. As can be seen in

Figure 4-12 the peak voltage across the rectifier is the capacitor voltage (positive) to the peak negative swing of the secondary voltage. In other words the PIV applied to a rectifier is substantially equal to the peak-to-peak voltage of the ac source. This is 2 × 1.4 × secondary rms voltage. In the case of a 6.3 volt source, the PIV is 2.8 × 6.3 = 17.64 volts. This requirement is of little importance in most low voltage power supplies since the minimum PIV is generally 50 volts, well above the 17.64 volts required from a 6.3 volt source or even the 35.28 volts from a 12.6 volt source.

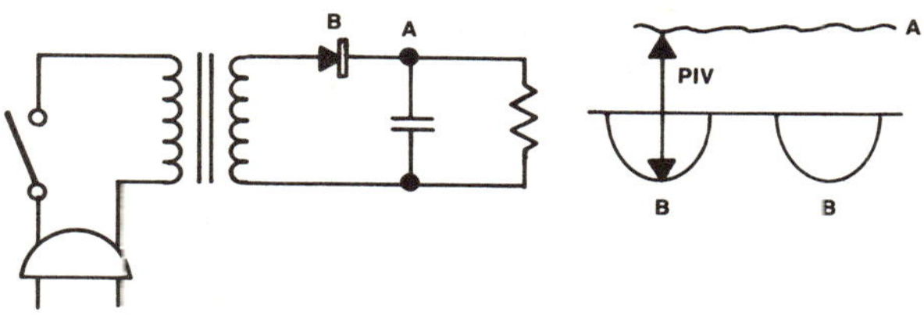

Figure 4-12—PIV Applied to a Rectifier

Two standard types of regulation which account for both line voltage changes and load changes are shunt regulators and series regulators. These regulators not only maintain constant output or load voltage but they also reduce ripple.

Shunt Regulators

In applications where the load current is substantially constant and where it does not exceed a predetermined value, the simple zener diode shunt regulator is applicable. As in all regulators of the shunt or series variety, the source voltage must be greater than the sink (output) voltage. Regulation consists in varying this difference to compensate for changes in source voltage and load current.

The basic zener diode shunt regulator circuit is shown in Figure 4-13. Several calculations are involved here. First, resistor R must be chosen so that the zener diode will not be overloaded. Now $P = (E_{IN} - E_{OUT})^2/R$. In the actual model built and tested, the Zener was rated at 1 watt but to be very conservative 0.3 watt was used in the calculations. The input voltage was 16 volts and the output (regulated) voltage was 5 volts so that R was calculated to be 183 ohms. 180 ohms was selected as the nearest standard value.

The power dissipation in R is equal to $(E_{IN} - E_{OUT})^2/R$ which came out as 0.67 watt. Again, to be conservative a 2-watt resistor was used.

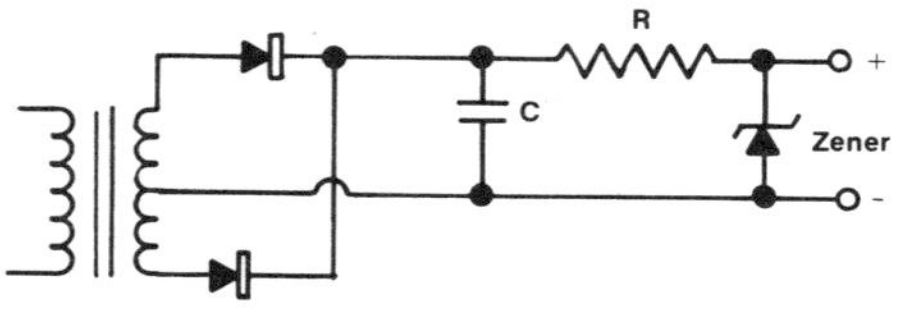

Figure 4-13—Zener Diode Shunt Regulator Circuit

PROJECT: Build a Dual Shunt Regulated Power Supply

Starting with two 12.6 volt 300 ma step-down transformers, four 300 mfd 35-volt capacitors, two 2-watt 180-ohm resistors, two

Figure 4-14—Dual Shunt Regulated Power Supply Completed

1-watt 5-volt zener diodes, switches, binding posts and a 6 × 3½ × 1⅞ inch phenolic box, build a dual 5-volt, shunt-regulated power supply using two circuits like the circuit of Figure 4-13.

This dual 5-volt power supply is just the thing for experiments with TTL integrated circuits which are rated at 5 volts. It should find many other uses around the experimenter's laboratory.

The finished dual 5-volt shunt-regulated power supply model is shown in Figure 4-14 and the inside assembly and wiring is shown in Figure 4-15.

While the simple shunt regulator just described is adequate for many applications, finer regulation and still further reduced ripple can be achieved by adding amplification and an active transistor as the shunt device.

After the dual-regulated power supply has been completed, test it and measure its characteristics. In this way, one can determine what functions it can perform.

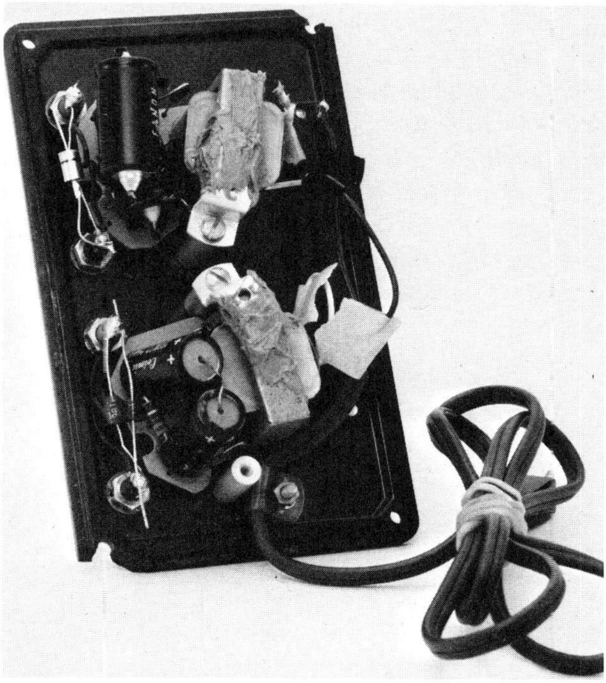

Figure 4-15—Inside Dual Shunt Regulated Power Supply

Series Regulation

A controlled resistance in the form of a transistor in series with a raw dc source is the basis of the series regulator. The series transistor in the power supply described in Chapter 1 uses a transistor as a variable resistor. This circuit becomes a series regulator when the transistor is controlled by an error voltage, generated by comparing the output voltage with a known reference voltage.

A basic series regulator circuit is shown in Figure 4-16. The unregulated voltage source is represented as a battery, E, for simplifying the diagram. This, of course, would be the transformer-rectifier-capacitor combination comprising an unregulated source. The output terminals are shown labeled plus and minus and, across which is connected the load resistor. Connected in series with the positive line is the pass transistor, Q_1, an NPN power transistor with its collector connected to the positive side of the unregulated source and its emitter connected to the positive output terminal. Control of the effective resistance of this pass transistor is accomplished by comparing a portion (or all) of the output voltage with the voltage of a reference zener diode, D. Transistor Q_2 receives a base voltage from potentiomenter R_3 and an emitter voltage from across zener diode D. The collector of transistor Q_2 is connected to the base of pass transistor Q_1. Drive current for the base of Q_1 is supplied through resistor R_1. If Q_2 were not present, Q_1 would be driven to saturation by current flowing through resistor R_1 and there would be only a very small voltage drop between the positive

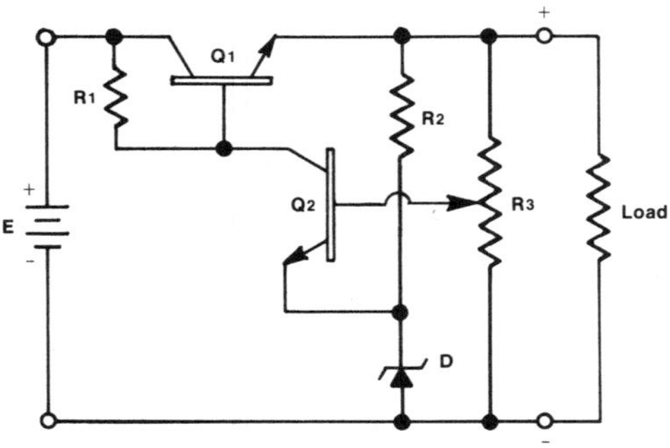

Figure 4-16—Basic Series Regulator Circuit

terminal of the unregulated source and the positive output terminal. However, with Q_2 present, if the output voltage rises above its predetermined value, in other words, if the base of Q_2 goes positive with respect to the reference at its emitter, Q_2 conducts and pulls down the voltage on the base of pass transistor Q_1 causing it to conduct less strongly. Equilibrium is established when the output voltage is reduced so that the base voltage of Q_2 equals the emitter voltage (plus the usual base-to-emitter drop). A change in unregulated input voltage or load voltage will cause the circuit to reestablish the predetermined output voltage by drawing more or less current through resistor R_1.

The regulation of any regulated power supply depends on the degree of amplification in the comparison (error) amplifier. It is therefore only natural to use readily obtainable, inexpensive and high gain operational amplifiers such as the 741 as shown in the simplified schematic of Figure 4-17. There is, however, the practical disadvantage that an auxiliary power supply will generally be required to supply the positive and negative bias to the amplifier.

The operational amplifier of Figure 4-17 is actually acting as a differential or comparison amplifier. Such amplifiers are available

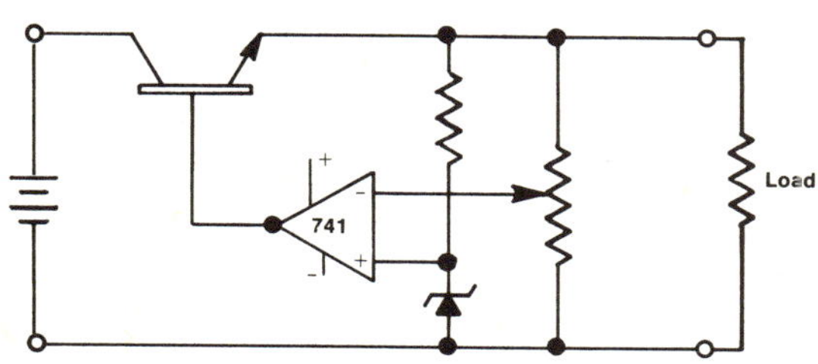

Figure 4-17—741 Operational Amplifier

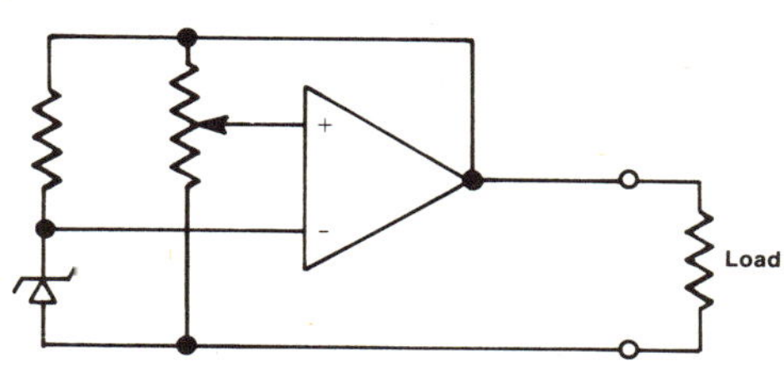

Figure 4-17A—Simplified Circuit for Op Amp

with varying output capabilities. As such, the pass transistor can be considered as an internal part of the amplifier and the simplified circuit of Figure 4-17A results.

Another possible configuration for a power supply is the more conventional operational amplifier circuit shown in Figure 4-18. The reference in this circuit is shown as battery E_R which can actually be a battery (a mercury battery is often used as a reference) or it may be a zener diode as above. In this circuit, the load voltage E_L is equal to $R_{VC}/R_R \times E_R$. Thus R_{VC} can be made variable and acts as a convenient output voltage control. An important advantage of this circuit is that the output can be varied down to zero, not possible with the comparison type circuits shown above. In those circuits the output voltage cannot be varied to a value below the reference voltage.

PROJECT: Build Any One or More of the Above Power Supplies.

The comparison types are useful where constant (or adjustable within limits) voltage is required. These conditions generally exist for powering specific pieces of equipment.

DESIGNING A LABORATORY POWER SUPPLY

In a very simplified form, the circuit of Figure 4-18 will be used as the basis for laboratory type power supplies. Since the laboratory supply must be closely regulated and must be variable in output voltage from its maximum capability down to zero, the above is a good starting point.

Since operational amplifiers generally have limited output capabilities they are unsuited for direct use as laboratory power

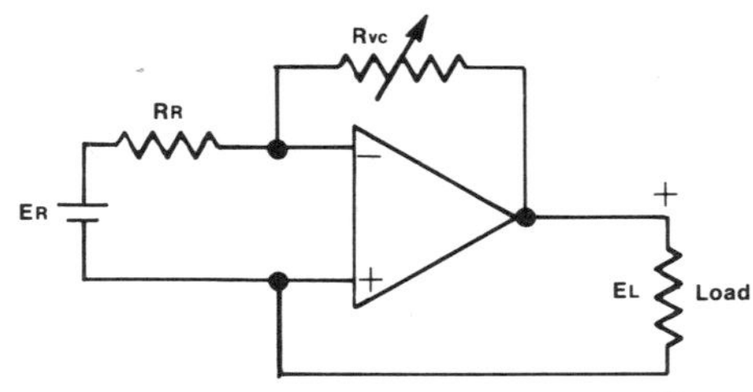

Figure 4-18—Basic Circuit for Laboratory Power Supplies

supplies. For example, suppose we rate our laboratory power supply at 0-15 volts at 0-1 ampere. The 741 operational amplifier, for example, would be entirely inadequate.

To provide more current, an emitter-follower (power transistor) as shown in Figure 4-19 can be added. There are, however, two limitations in this circuit. First of all, it is an emitter follower, which means that the base voltage must be equal to the emitter or load voltage (plus the emitter-to-base conduction voltage) which may tax the output of the operational amplifier. The second is that the power transistor requires drive current which again may exceed the capability of the operational amplifier.

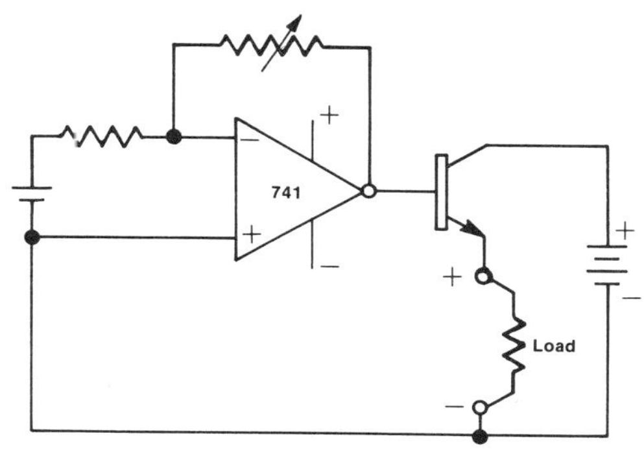

Figure 4-19—Emitter-Follower Added to Power Supply Circuit

Figure 4-20 is a simplified schematic circuit in which both of the above stated limitations have been overcome. It will be the basis for the laboratory power supply project to be described in detail below. The drive current problem has been solved by using a Darlington pair transistor as the pass transistor. The drive current for a 1-ampere load current is only a few milliamperes, well within the capability of the 741 op amp. The load has been transferred from the emitter to the collector. In this way, only a small base voltage on the Darlington transistor is required for full 15 or more output volts.

Figure 4-20 is divided into basic functional areas. The reference voltage and operational amplifier are designated as the control unit. By using a small auxiliary power supply, this unit provides a control unit of considerable versatility. It can be combined with a wide range of pass transistors and unregulated sources to provide power supplies capable of a wide range of output voltage and current which all go to zero and are well regulated.

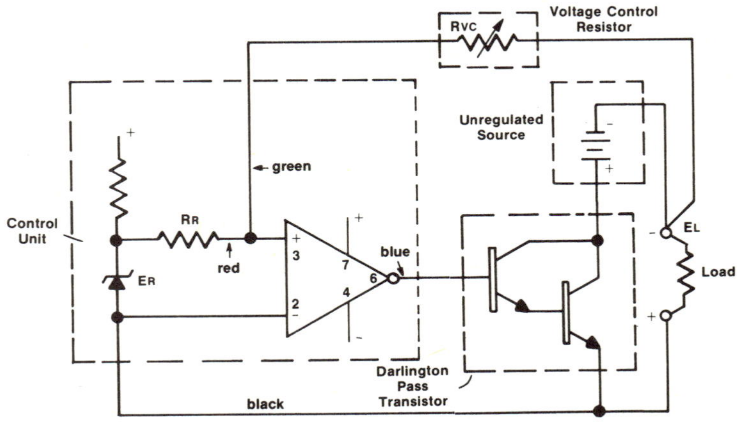

Figure 4-20—Power Supply Circuit with Darlington Pair

Figure 4-21 shows this control unit in more detail. In order to simplify assembly, the leads are color coded.

Returning to Figure 4-20, the Darlington pass transistor is shown in a dotted box as are the two remaining essential components, the unregulated source and the voltage control resistor. To simplify computations, if E_R/R_R is made equal to 1 milliampere, the output voltage will be 1 volt per thousand ohms of the voltage control resistor. Thus, for a power supply to provide a maximum output voltage of 15 volts, the voltage control resistor would have a maximum resistance of 15,000 ohms (15K).

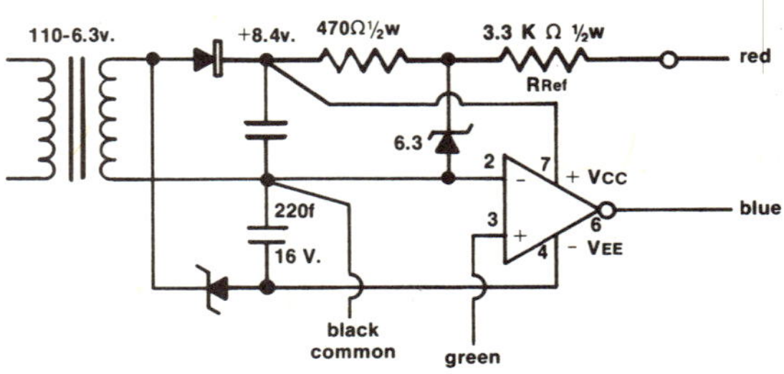

Figure 4-21—Control Unit of Power Supply

PROJECT: Build a 15 volt 1 ampere Laboratory Power Supply

The regulator using the circuit detailed in Figure 4-21 was built on a perf board as shown in Figure 4-22 and wired as shown in Figure 4-23. The unregulated source was wired as shown in Figure 4-24. These parts were mounted on the bottom of the phenolic case and located so that they would not interfer with the meters and other parts mounted on the panel. See Figure 4-27 for ideas regarding the arrangement. The following is a list of parts used.

Parts List for Laboratory Power Supply

Item	Radio Shack Part No.
6.3 volt line transformer	273-1384
12.6 volt line transformer	273-1505
220 mfd 16 v. capacitors	272-1006
470 ohm ½ watt resistor	271-019
3.3K ½ watt resistor	271-028
10K linear potentiometer	271-1715
741 operational amplifier	276-038
Binding posts	274-661
on/off switch	275-602
0-15 volt meter	22-054
0.1 ma meter	22-052
Heat sink	
0.17 ohm current meter shunt	
Silicon rectifiers (2)	276-1141
3000 mfd 30 v. capacitors	purchased from surplus
Darlington pass transistor	

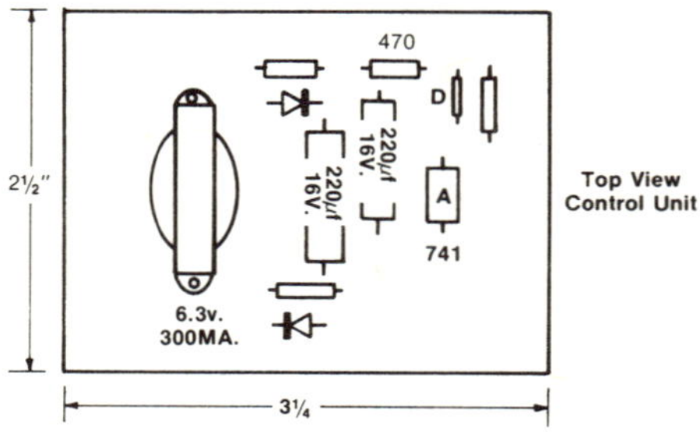

Figure 4-22—Lab Power Supply Control Unit Completed

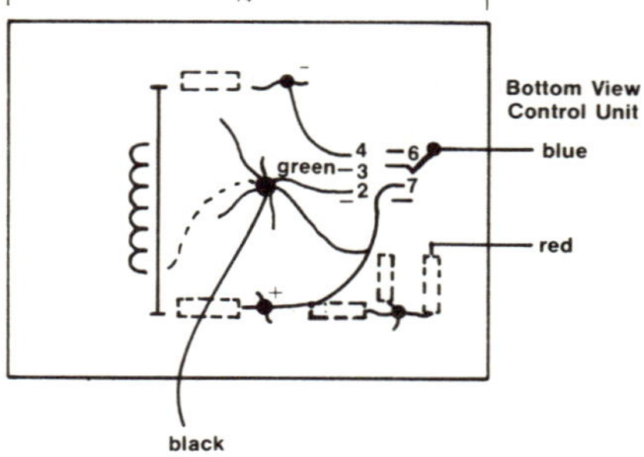

Figure 4-23—Wiring of Lab Power Supply Control Unit

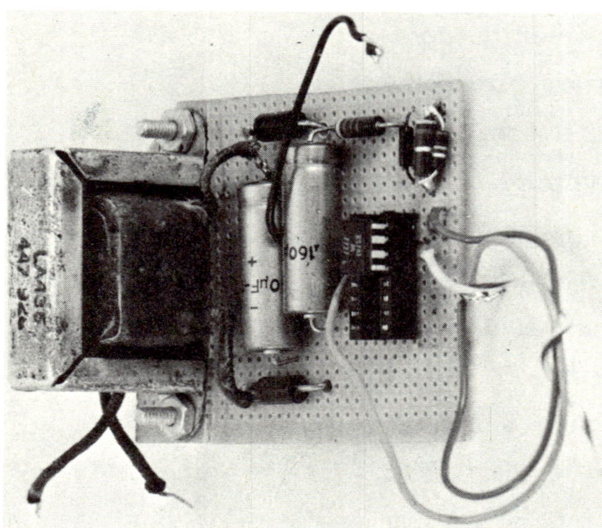

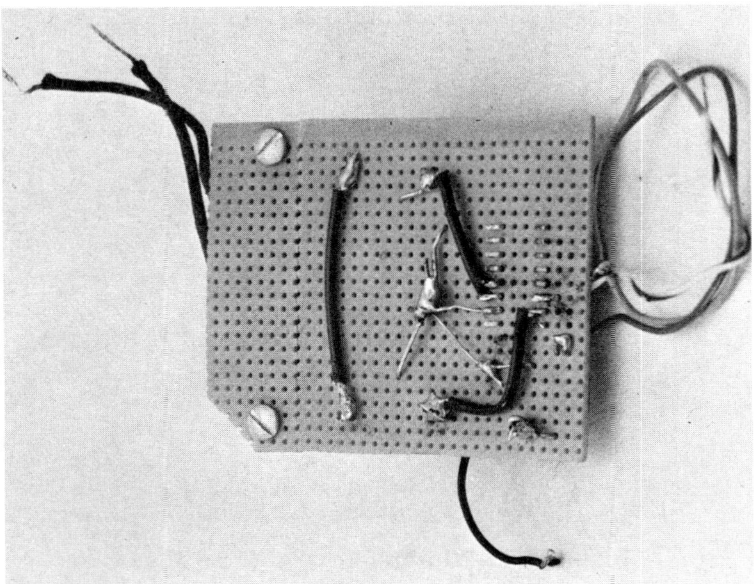

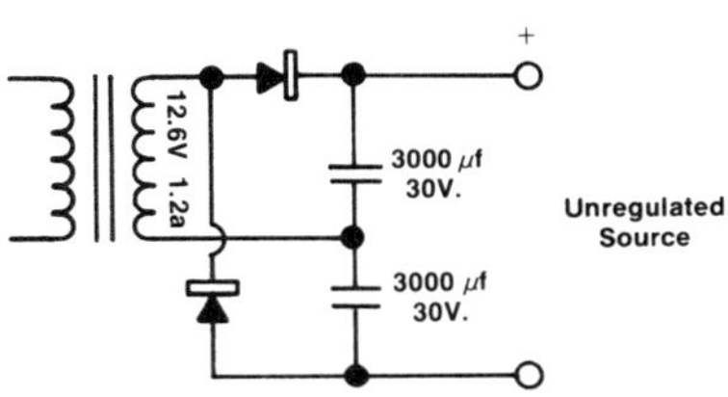

Figure 4-24—Wiring of Unregulated Source

The box panel was drilled as shown in Figure 4-25. The template furnished with the meters was used to locate the meter holes. The finished laboratory power supply is shown in Figure 4-26. The control unit unregulated source, voltage control potentiometer,

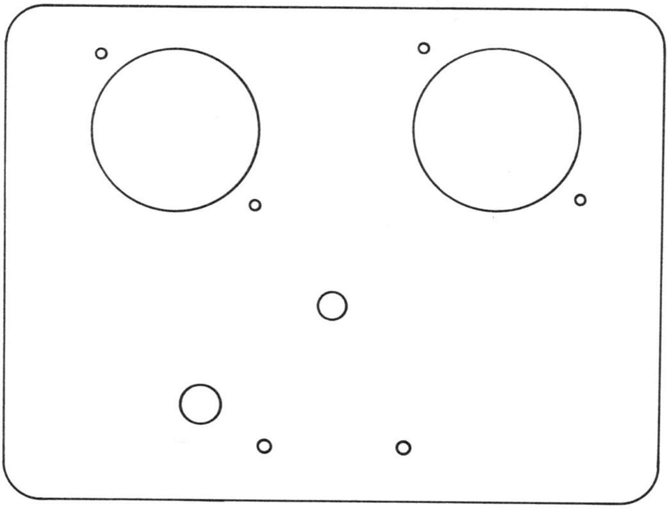

Figure 4-25—Bottom View of Front Panel

Figure 4-26—Laboratory Power Supply Completed

pass transistor, meters and output binding posts were interconnected as shown in Figure 4-20.

Note: The original Darlington and heat sink as shown in Figure 4-26 was later changed to the Radio Shack 276-2042 and mounted on a heat sink outside the original box on a large heat sink.

Once your equipment is operating as it should, proceed with performance testing as follows:

1. Line voltage regulation. Change the line voltage 10 volts and note the change in regulated output voltage. Calculate the percentage change in output voltage.

2. Load voltage regulation. Change the load from zero to full-rated load and note voltage change. Express the change in percentage of output voltage.

3. Peak-to-peak ripple at full load. Measure and express in millivolts.

These three tests will quite adequately define the quality of your power supply. It would be interesting to compare your figures with published figures of commercial power supplies.

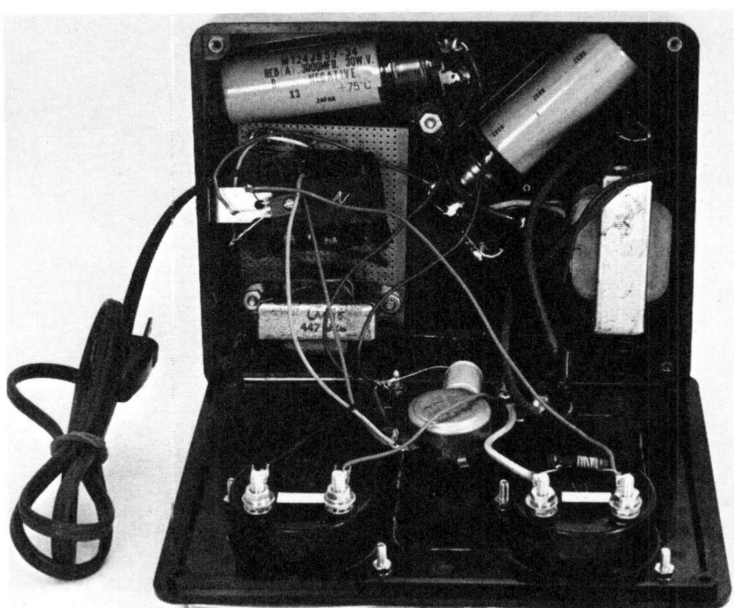

Figure 4-27—Laboratory Power Supply Inside

BUILDING AN AUTOMATIC CROSS-OVER POWER SUPPLY

Current regulation in a power supply can be achieved in much the same way that voltage regulation is attained. However, current regulation is generally not as precise as voltage regulation, mainly due to the fact that, for practical reasons, a smaller sample is used for control.

Figure 4-28 is a simplified schematic diagram of an automatic cross-over (E/I) regulated power supply. The auxiliary power supply is the same as that shown in Figure 4-21. The unregulated power supply is shown as a battery. The current is sampled by resistor R_I, is referenced by the zener diode, supplied with a negative voltage and is controlled by R_{IC}. The second operational amplifier for the current control is OR-gated with the voltage amplifier to the base of the driver transistor. The current control circuit can also be used as a current limiter.

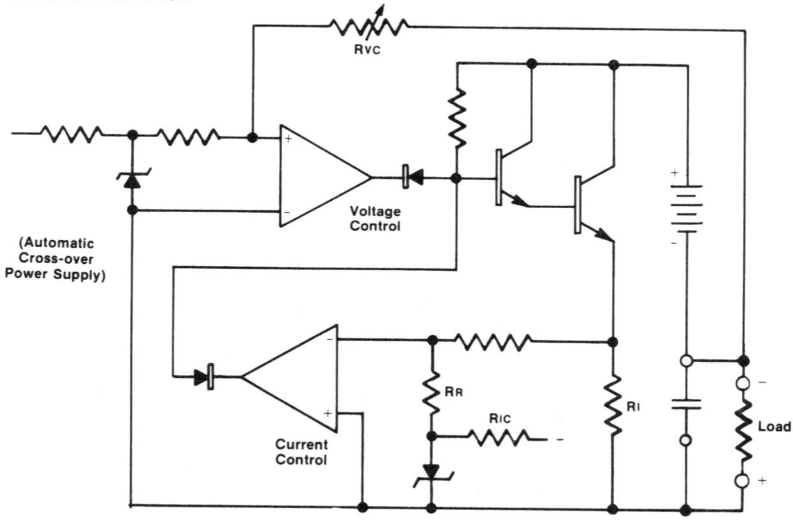

Figure 4-28—Circuit of Automatic Cross-Over Regulated Power Supply

BUILDING A DUAL TRACKING POWER SUPPLY

A dual tracking power supply is one which provides adjustable but equal positive and negative regulated voltages. A single reference and adjustable voltage control are used, as in the case of a single-ended regulated supply. A second amplifier is slaved to the first power supply to provide the second polarity as shown in Figure 4-29. Here, two amplifiers and two driver/power transistor combinations are used, one for each polarity.

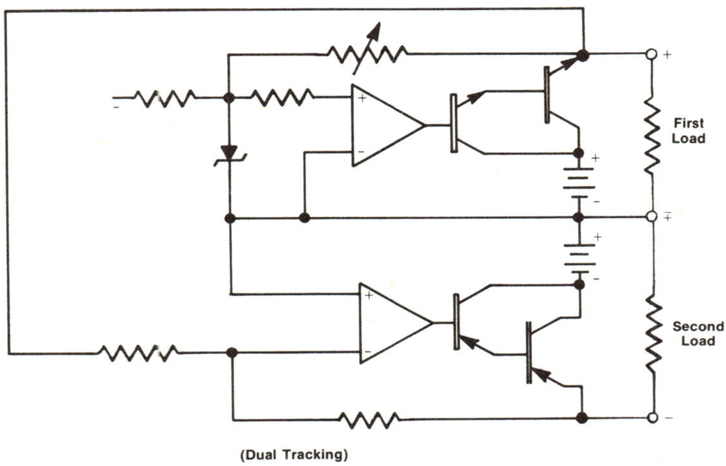

Figure 4-29—Second Amplifier Slaved to First Amplifier

In the regulating circuits up to this point, the output or load is placed in the collector circuit so that only a small driving voltage is required to the pass transistor. In Figure 4-19 a switch has been made and the output or load circuits are effectively in the emitter circuits, constituting emitter followers. In this circuit the op amps must have a compliance equal to the output voltage plus the base-to-emitter drop of the driver plus the base-to-emitter drop of the pass transistors. This circuit has the advantage of using a common point between the two sources of unregulated voltage. For low voltage supplies, the compliance requirement is readily met. One difference between the two types of circuits is the phase. The collector-loaded circuit inverts the phase in the pass stage so that degenerative feedback is made to the non-inverting input of the op amp. In the emitter-follower type of circuit there is no inversion in the pass stage so that degenerative feedback goes to the inverting input of the op amp.

HOW TO FAST-PROGRAM A UNIPOLAR POWER SUPPLY

The usual unipolar dc power supply is provided with a large bypass capacitor across its output. When increased voltage is called for, the pass transistor resistance is sharply decreased, resulting in a fast charging of this capacitor. However, when decreased voltage is called for, the capacitor has only the load resistance path to discharge it. This often results in a situation where output voltage can be increased rapidly but can only be decreased very slowly. There are two ways to equalize the rates of increase and decrease in the output voltage. One way is to drastically reduce the capacitance across the load and the other is to provide a dynamic control of the decreasing voltage as well as the increasing. Decreasing the output capacitance can lead to instability and other problems.

Figure 4-30 is a simplified schematic circuit diagram of a dynamic up-down series/shunt regulator for unipolar power supplies. Much the same considerations of current capacity and cooling requirements apply to the shunt regulator as to the series pass transistor. However, this type of power supply is suitable for use in automatic testing systems and in other applications where fast down-programming is required as well as fast up-programming.

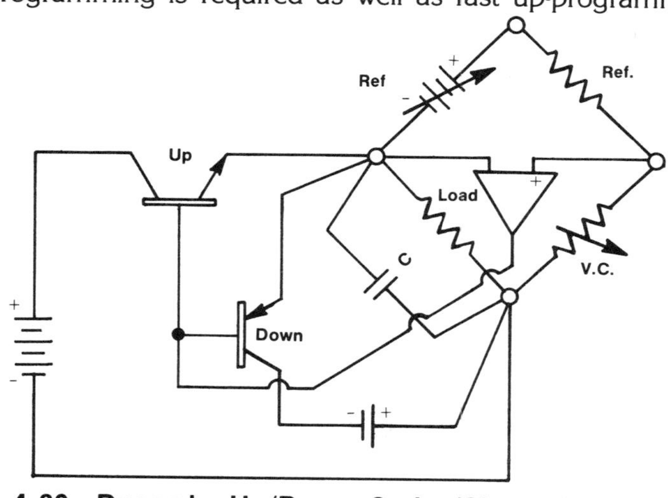

Figure 4-30—Dynamic Up/Down Series/Shunt Regulator for Unipolar Power Supplies

BIPOLAR POWER SUPPLIES

A bipolar power supply is essentially a power amplifier which follows direct current as well as alternating current commands. Short circuit protection is important in either application. The main difference is that the slewing rate of power supplies need not be as great as in the case of audio amplifiers. See Chapter 10 for power amplifiers which can double as bipolar power supplies.

REMOTE CONTROL HEAD FOR REGULATED POWER SUPPLIES

A remote control box provides a handy addition to bridge controlled power supplies. This is a unit carrying output current with current- and volt-meters and a voltage control rheostat as shown in Figure 4-31. This unit is housed in a 3½ × 6¼ plastic box and forms a convenient link between a power supply and an experimental set-up. The power supply can be on a shelf out of the way, while the remote control box, being quite small, can be conveniently near the experimental set-up on the bench. The schematic arrangement is shown in Figure 4-32. With the box in plain view along side the experiment, a time-saving and convenient arrangement is provided. Figure 4-33 shows the inside of the remote control head.

Figure 4-31—Remote Control Head

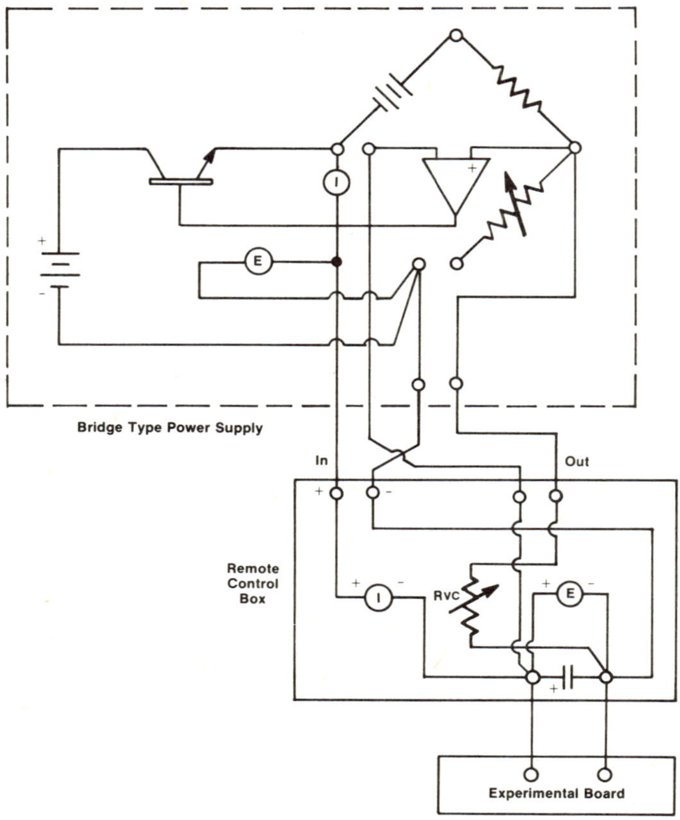

Figure 4-32—Schematic of Regulated Power Supply with Remote Control

Figure 4-33—Inside of Remote Control Head

5

Projects and Experiments with Silicon Controlled Rectifiers and Triacs

The silicon controlled rectifier (SCR) evolved from the earlier four layer diode which, as implied by its name, was a P-N-P-N combination. Once the SCR was fully conceived, its uses and applications spread rapidly. The SCR in simplest terms is a solid state current or voltage controlled switch. In order to more readily understand the SCR, it may be simulated by two transistors, one PNP and the other NPN as shown by the two most commonly used symbols in Figure 5-1.

As shown, the upper PNP transistor is connected to the lower NPN transistor base-collector and emitter-base respectively. This will be seen to form a four-layer device P-N-P-N with two of the N and two of the P elements connected together. If the anode is made positive and the cathode negative, current will be blocked at the common N-P junctions. Similarly, if the cathode is made positive and the anode negative, current will be blocked at these same junctions. Thus, the device blocks from both directions. This blocking characteristic must be considered in using the device. If breakdown takes place in either direction under operating conditions, control of the device is lost and it may even be destroyed. This breakdown voltage is called "stand-off voltage" and is generally specified for any given SCR.

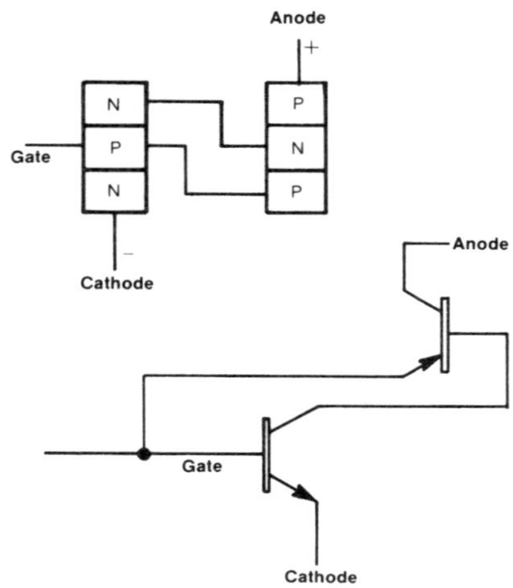

Figure 5-1—Junction of PNP and NPN to Simulate SCR

Now, in order to control the SCR, a gate connection is provided as shown, connected to the base of the lower, or NPN, transistor. When a voltage or current (positive with respect to the cathode) is applied to the gate, current flows across the lower P-N junction and, by transistor action, induces current to flow across the upper N-P junction. However, the connection to the N-P junction of the upper transistor (PNP) causes a regenerative action and a critical point will be reached when an avalanche action takes place and the device breaks down. The drop from anode to cathode essentially becomes that of two P-N diodes in series. The breakdown voltage is specified as gate trigger voltage or current.

One further characteristic of the SCR is critical; the holding current. This is the minimum current from anode to cathode which will maintain conduction when the gate voltage or current is removed or returned to zero.

In order to turn an SCR off it is generally necessary to reduce the anode-to-cathode current to zero, or at least below the holding current. This is usually done by reducing the anode-to-cathode voltage to zero, either by opening the anode circuit or by shorting anode to cathode. With alternating current input, turnoff takes place automatically as the input voltage passes through zero.

There are a number of secondary effects to which the SCR responds because of capacitances of the diode junctions. For example, a very sharp increase in anode voltage can turn an SCR on, although a fast transit down through zero may not turn it off.

The conventional symbol for the SCR is shown in Figure 5-2.

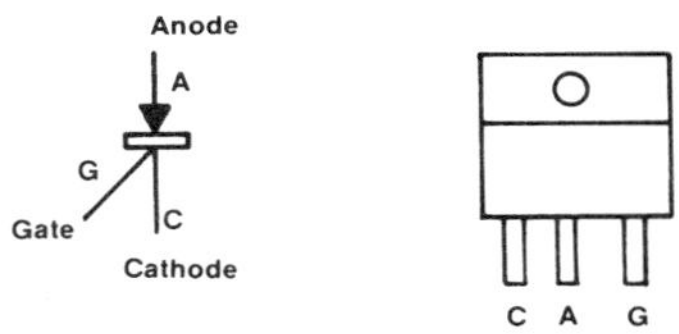

Figure 5-2—Conventional Symbol for SCR

SYNTHESIZING AN SCR WITH TRANSISTORS

Select two general purpose silicon transistors of moderate power rating, one PNP and the other NPN. Connect them as shown in Figure 5-3. The R-C shunt across the simulated SCR is made to

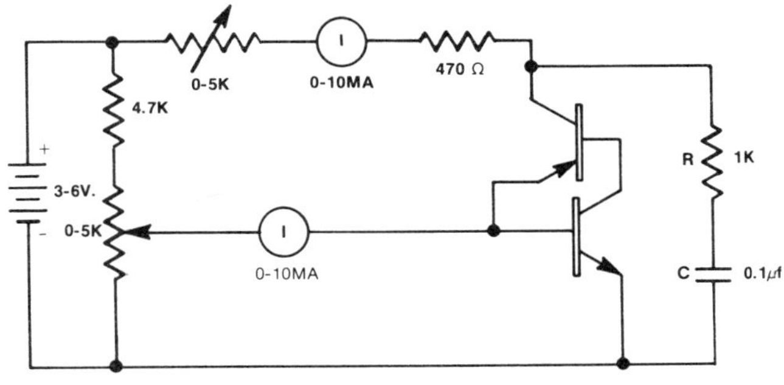

Figure 5-3—Synthesis of SCR with Transistors

prevent triggering the SCR through its internal capacitances if the anode voltage is suddenly applied.

Now, starting with zero gate voltage, check the anode current. If it is not zero, turn off the anode voltage and start again. If the SCR conducts at zero gate current, the transistors probably have too much internal leakage and will not act like an SCR. You should therefore try other transistors.

Again, with transistors which do not conduct at zero gate current, increase the gate current until anode current flows. Note the turn-on current to the gate.

Turn down the gate current. If the anode current drops, the anode current is less than the holding current. Progressively increase the anode current by decreasing the value of the anode series resistor until the anode current holds when the gate current is decreased to zero. Thus, the value of anode current which just holds is the anode holding current.

It will be noted that the device will probably conduct at a very low gate current and hold at a low anode current. These are characteristics of the so-called sensitive gate SCRs. While this simulated SCR helps understand SCR operation, actual SCRs have been developed to have characteristics much more useful in most practical applications.

EXPERIMENTS IN TURNING SCRS ON AND OFF

First, set up the circuit for determining turn-on and holding currents as shown in Figure 5-4. The values to be found using standard SCRs will be found considerably different from those of the sensitive gate variety. It is interesting to plot anode current, I_A, versus gate voltage, E_g, and current, I_g, as shown typically in Figure 5-5.

Some applications have been made of this sharp trigger-like anode current characteristic in devices operated below holding current. With a low gate input impedance there is very little hysteresis in this switching characteristic. Adding series resistance to the gate increases the hysteresis (falling current does not track rising current).

While it is sometimes possible to turn off an SCR with a negative-going gate pulse, it is generally assumed that turn-off will be accomplished by opening the anode circuit or by shorting the anode to the cathode. Instead of an actual conductive short, a capacitor may be used to simulate a short.

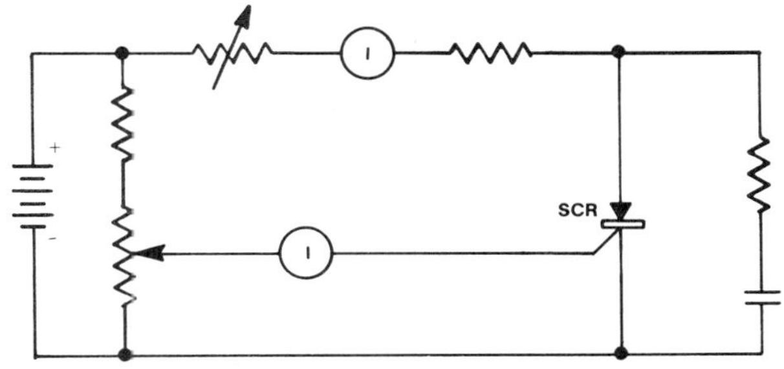

Figure 5-4—Circuit for Determining Turn-On and Holding Currents

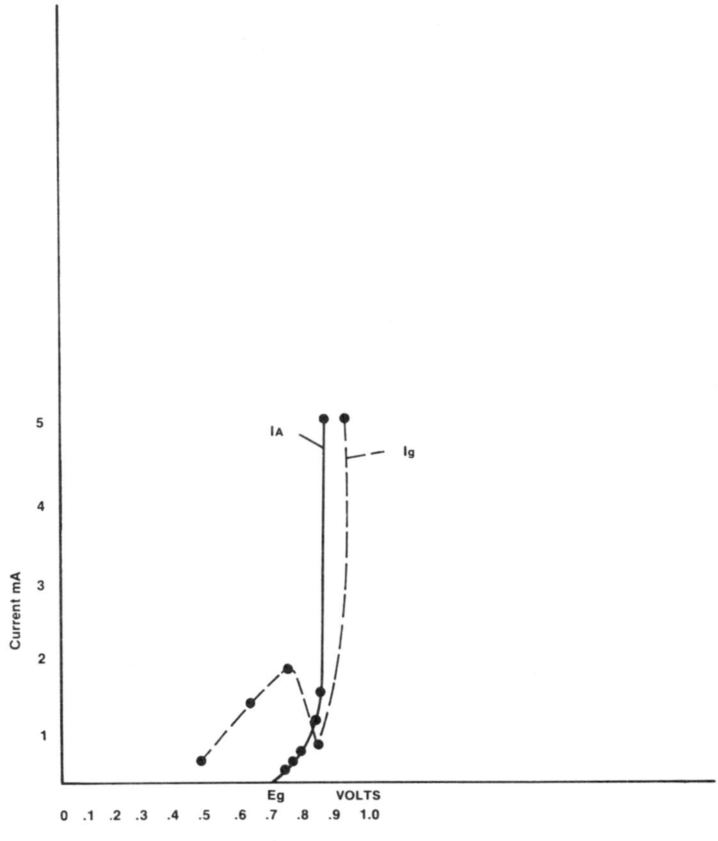

Figure 5-5—Anode Current Versus Gate Voltage and Current

Experiment: Set up the circuit of Figure 5-6 and determine the value of C, which turns the SCR off when suddenly switched from anode to cathode. How, then, is value of turn-off capacitor affected by the current being passed by the SCR at the time of turn-off?

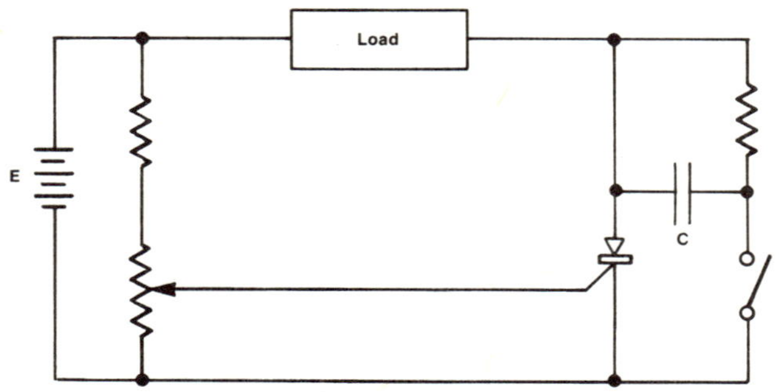

Figure 5-6—Experimental SCR Circuit

SIMULATING A TRIAC WITH SCRS

The SCR is a current- or voltage-controllable switch which conducts unilaterally when broken down or triggered. When operated on AC it conducts only during the positive half cycle of the AC input and goes off automatically as the anode voltage passes through zero. It has many applications as a half-wave conductor such as, for example, a half-wave controlled rectifier. Two SCRs connected in a rectifier bridge will control a full-wave rectifier, as will be described below.

A related device, which in itself operates as a full-wave device, is the triac. A triac may be simulated by means of two SCRs as shown in Figure 5-7. The symbol for the triac is shown in Figure 5-8. A device which is often used to control a triac is the diac, which is a two-way conducting trigger diode. The diac symbol shown in Figure 5-8 is similar to that of the triac, but is without a gate connection and without polarity designation. While the traic is a two-way conducting device, it is so constructed that the M1 terminal is connected to the common side of the circuit, as will be obvious from application circuits described below.

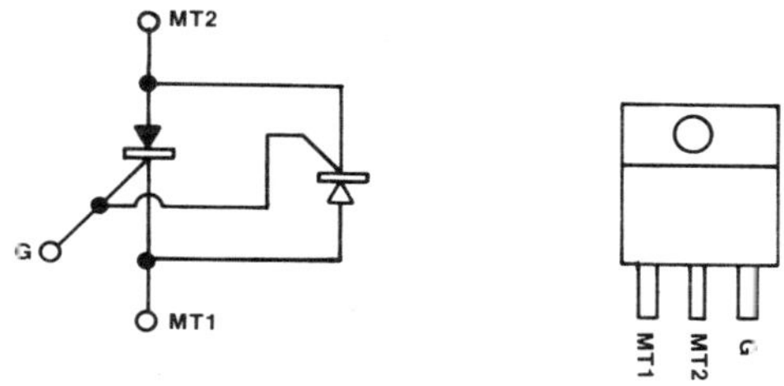

Figure 5-7—Simulated Triac Using Two SCRs

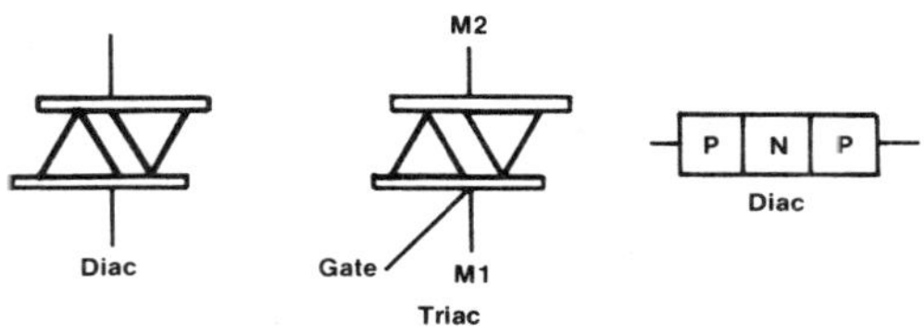

Figure 5-8—Symbols for Diacs and Triacs

PROJECT: Make Your Own Light Dimmer

Since the triac provides full-wave control, it will provide full-range dimming. A simple light dimmer, suitable for controlling a 100-watt lamp can be built using the circuit shown in Figure 5-9. This was put together in a small utility case, using the parts listed on Figure 5-10, which is a photograph of the internal arrangement and finished unit.

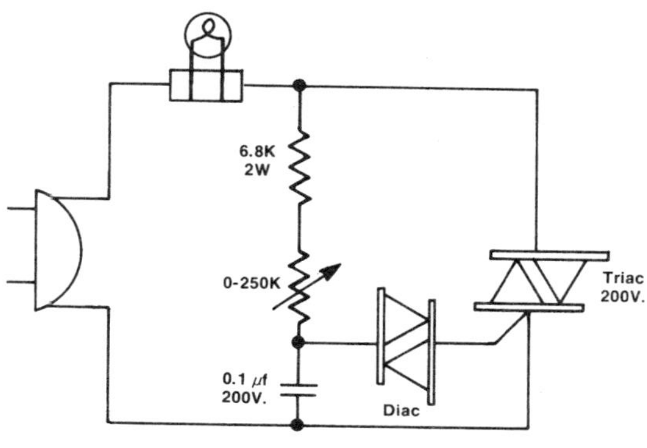

Figure 5-9—Dimmer Switch Schematic

Figure 5-10—Light Dimmer Inside and Completed

The metal tab of the triac was insulated from its internal elements, therefore no insulation between it and the potentiometer case to which it was epoxied seemed necessary. Since the peak voltage of a 120-volt source is 170 volts, a triac rated to at least 200 volts should be used.

In locations where interference-sensitive equipment, such as AM radios, is located, the dimmer may produce objectionable radio frequency interference (RFI). This can be effectively eliminated by means of a simple RFI filter, as shown in Figure 5-11. All that is required to be added is a 100-microhenry choke and a 0.1 microfarad capacitor. The choke is readily made using a ⅜ inch dowel or an Allen Bradley 2-watt resistor as a form. Details of the 100-microhenry choke are also shown in Figure 5-11. Keep leads, particularly to the capacitor, as short as possible.

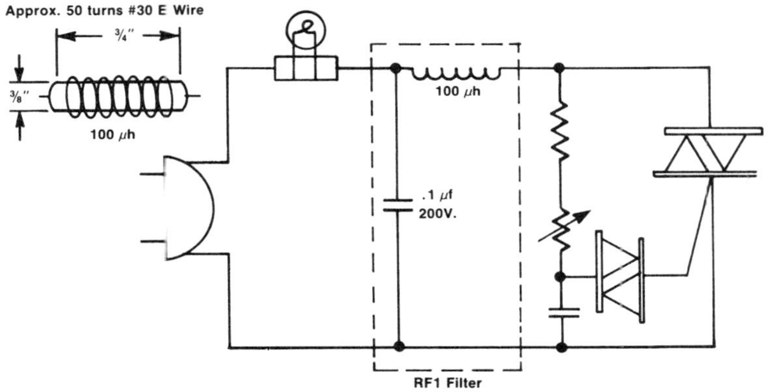

Figure 5-11—Addition of RFI Filter to Dimmer

REGULATING A POWER SUPPLY WITH SCRS

SCRs have been widely used in regulated power supplies. When used in the ac circuits, either before or just after the power transformer, they provide means for regulating the dc output. When compared with series or shunt regulators used in the dc circuits they are generally slower and, more importantly, do more to increase ripple on the output than to suppress it. However, they provide very efficient regulation and are quite satisfactory where very low ripple is not important.

Figure 5-12 shows a simple regulator for a low-voltage power supply using a single SCR.

Experiments: Measure the output voltage range (note the output voltage cannot be adjusted to voltages substantially below the voltage of the zener reference voltage). Why is the auxiliary power source for the zener and amplifier transistor necessary? Measure regulation. Measure ripple versus load current for various values of filter capacitor C. Does the SCR require a heat sink when maximum output is being supplied? What conditions can cause the amplifier transistor to fail?

It will be noted that the regulator of Figure 5-12 controls from the maximum available rectified voltage down only to a voltage equal to the zener voltage plus V_{EB} of the amplifier transistor. With some simple additions regulation can be extended down to V_{EB} or about 0.5 volt minimum as described below.

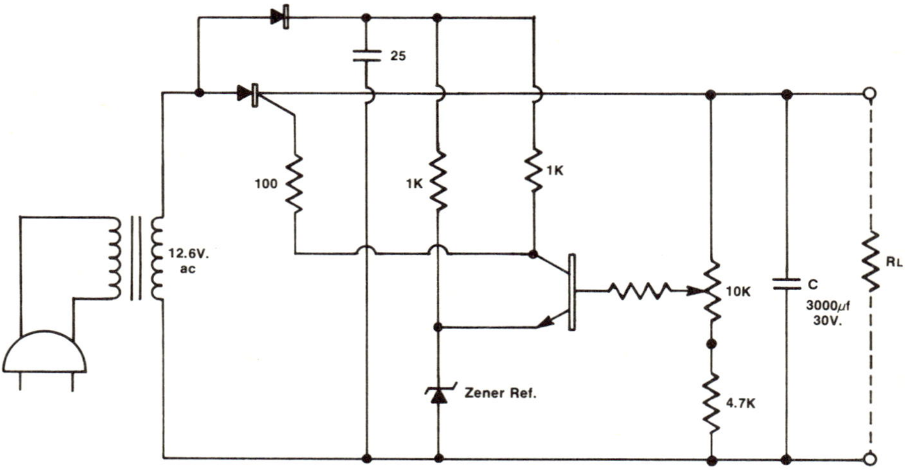

Figure 5-12—Regulator for Low-Voltage Power Supply

Wide Range SCR Regulator

With the addition of a simple auxiliary dc source and a rearrangement of components as shown in Figure 5-13, a wide-range regulated power supply is provided. Such a power supply is generally useful in an experimental set-up.

Experiments: Perform the same experiments as outlined for the circuit of Figure 5-12. It is suggested that this latter power supply be packaged for use as an experimental source of regulated dc voltage. The 10K variable resistor can be calibrated and marked in volts output. A voltmeter and an output current meter can be added for even greater usefulness.

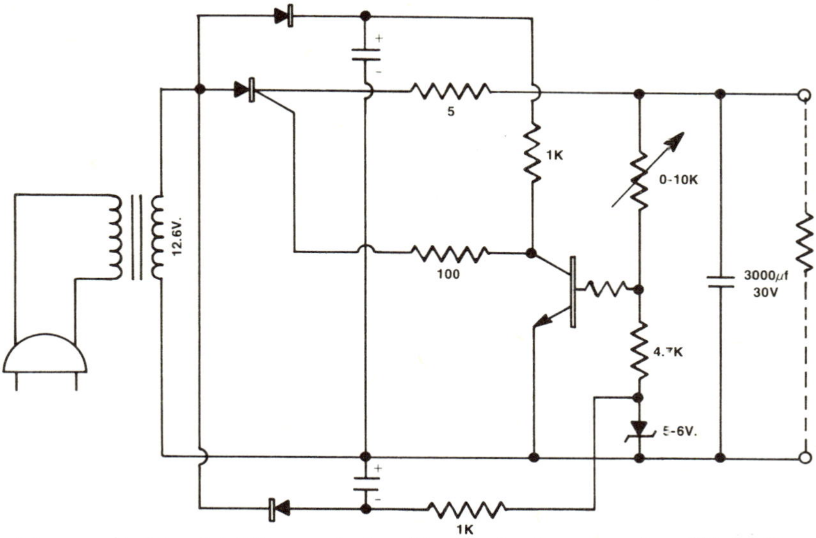

Figure 5-13—Addition of Auxiliary dc Source for Wide-Range Regulated Power Supply

COMPARING SCRS AND TRIACS AS CONTROLLERS

As has been described above, SCRs and triacs are triggerable avalanche devices, intended for use as controlled ac switches. The SCR conducts in only one direction, so that when used with an ac input, it can pass only alternate half-cycles of the ac wave. Thus, only the half-cycle power is available in the output. For example, a 100-watt lamp, dimmed by a dimmer using an SCR, will only have a maximum brilliance of a 50-watt lamp.

The triac, on the other hand, controls both cycles of the ac input and hence can provide full output to an ac-operated device. Where maximum available power is not required, the SCR may be found to be a suitable control device.

Comparing SCRs and triacs in another way, the SCR is suitable as a combined rectifier and controller, as in a regulated power supply, whereas the triac does not rectify and is therefore useful only as an ac controller.

TESTING SCRS, DIACS AND TRIACS

Testing components is an important part of experimental electronics. This is especially true in view of the uncertain condition of many and perhaps most of the components available to experimenters. However, in order that testing not be a burden and may be more of a learning process, only characteristics which are important for a given application need to be tested.

Perhaps the most important, and generally the easiest, of all tests are those used to identify solid state devices. For purposes of this section, the first tests to be described are those which will identify diacs, triacs and SCRs.

Diac Testing: The diac is a PNP combination like a PNP transistor, except that it has only two leads; both PN junctions are similar in that they provide symmetrical operation. The junctions provide a stand-off condition until avalanche breakdown takes place, after which they exhibit a negative resistance. Test for breakdown in both directions using an adjustable voltage and a current limiting resistor as shown in Figure 5-14. The voltmeter will show breakdown at the point at which voltage across the diac stops increasing and starts to decrease. Substituting an adjustable ac voltage for the dc and connecting an oscilloscope to observe voltage across the diac and current through it will result in the pattern shown for a good diac. Erratic behavior or a highly unsymmetrical pattern will show up a defective diac.

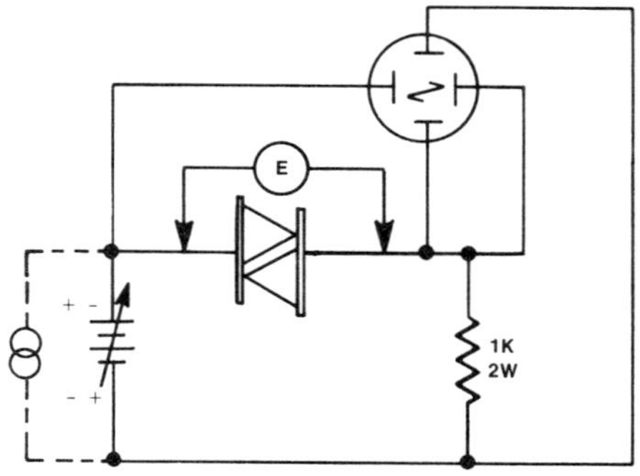

Figure 5-14—Circuit for Testing a Diac

Identifying SCRs and Triacs: The essential difference between SCRs and triacs is that the former are unipolar (breakdown in only one direction) and the latter are bipolar (breakdown in both directions). An ohmmeter test of an SCR will show conduction only from gate (G) to cathode (C) when the gate is made positive with respect to the cathode.

Triacs, on the other hand, show conduction from gate (G) to main terminal 1 (MT1) in either polarity.

Testing SCRs: Two significant tests of SCRs are easily made. First, stand-off voltage, the voltage which can be applied to the SCR before breakdown. If the SCR is to be operated in a 120-volt ac circuit, it must stand off a voltage in both directions greater than the peak of 120 volts or greater than 170 volts plus a margin of safety to provide toleration for spikes and elevated temperatures. It is usual to specify a stand-off voltage of 200 volts for 120-volt ac operation and 400 volts for 240-volt operation.

Figure 5-15 shows a typical test set-up with a table of test voltages, protective resistor values for 2-watt resistors and maximum current which can flow on breakdown of the SCR. Connections for forward stand-off voltage are shown. For reverse stand-off voltage, merely reverse the connections to the SCR.

Gate sensitivity is the second important test. SCRs are broadly classified as standard or sensitive gate. The SCR gate turns on at about 0.7 volts, but the current required may be several milliamperes for a standard gate or a fraction of a milliampere for a sensitive gate. Standard gate sensitivities are most generally used since they are more stable and less subject to small disturbances.

Gate sensitivity can be measured by means of the set-up shown in Figure 5-16. The input voltage E_1 should be set at some nominal

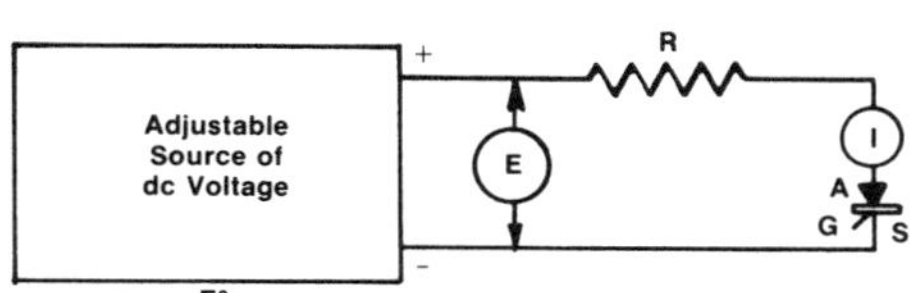

$R\ (2\ \text{Watt}) = \frac{E^2}{2}$

E Test	R	I
100V.	5K	20ma.
200V.	20K	10ma.
300V.	45K	6.7ma.
400V.	80K	5ma.
500V.	125K	4ma.

Figure 5-15—Circuit for Testing SCRs

voltage, say 100 to 200 volts. R_1 should be of sufficient value to protect the current meter I_1 when the SCR conducts.

A third characteristic, the holding current, can also be measured with the set-up of Figure 5-16. To find the holding current, the minimum current which will continue to flow when the gate voltage is removed, vary R_1 or E_1 until the anode current continues to flow when the gate current is reduced to zero. This is generally of the order of 5—15 milliamperes. As mentioned above, SCRs can be operated as on-off switching devices for dc if operated below the holding current.

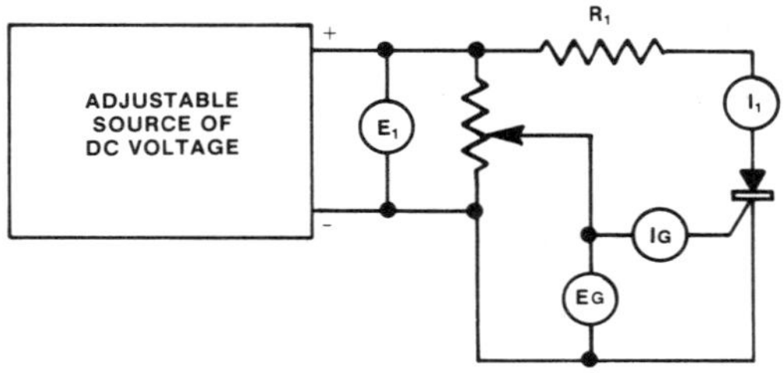

Figure 5-16—Circuit for Measuring Gate Sensitivity

The stand-off voltages and gate sensitivities of triacs are measured in the same way as set forth above for SCRs. The main difference is that a reversed polarity characteristic is to be measured by reversing the polarity of the anode and gate voltages. The common terminal is MT1 for both polarities and MT2 is the positive or negative other electrode. Figure 5-17 shows how the polarities are reversed but the gate voltage is always referred to main terminal 1 (MT1).

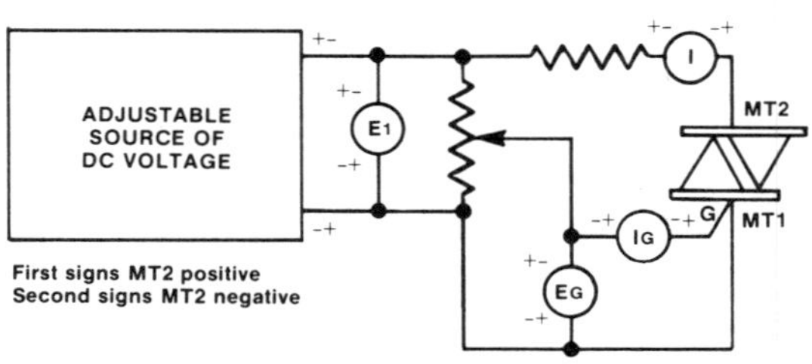

Figure 5-17—Circuit Showing Effects of Reversing Polarities

CONTROLLING MOTORS WITH SCRS

The most readily controlled motor is the widely used universal motor. An SCR can be used to control the speed of such a motor and will tend to regulate the speed under varying load conditions. Figure 5-18 shows a simple but effective circuit for speed control of universal motors.

Experiments: Observe the range of speeds over which the motor can be operated. How slowly can it be operated at a reasonably consistent speed? What effect does loading have on the motor speed? Does the SCR heat up? What practical uses can be made of this motor speed control?

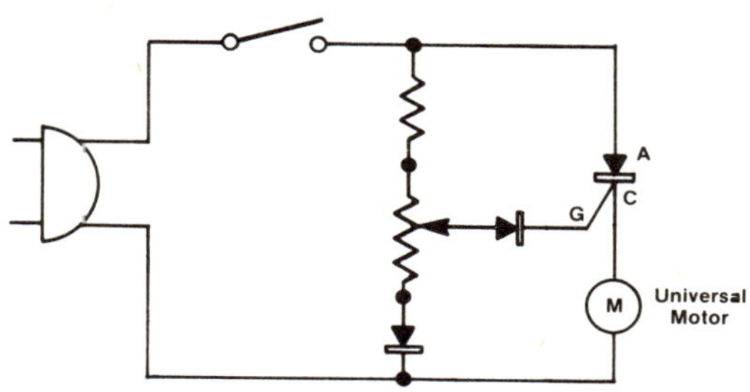

Figure 5-18—Circuit for Speed Control of Universal Motors

WHEN TO HEAT-SINK SCRS AND TRIACS

All but the very-low-power SCRs and triacs are constructed to enable effective heat-sinking. Studs and tabs are typical provisions. Most heat conducting provisions are electrically connected to SCR anodes or triac main terminals 2. Some triac tabs are insulated. In providing heat-sinking for these devices, the fact that they are generally electrically connected to a live terminal of the device must be provided for. Insulation such as mica or berillium oxide may be provided between the device and the heat-sink, or the heat-sink may be insulated from possible contact with human operators or external contacts.

SCR and triac controllers are relatively very efficient, for example, when compared with series resistive devices such as rheostats or series regulating transistors. Furthermore, maximum dissipation will generally be at minimum control in contrast with the series devices. They are switching devices which operate off or on in saturation.

Experiments: Set up a lamp dimmer as shown in Figure 5-9. Run with lamp at full brilliance for 10 minutes (no heat sink for SCR). Turn-off and feel the SCR. Repeat adding lamps in parallel until SCR gets hot. It will become too hot to touch at about 140° F. Since silicon devices are generally rated at operating temperatures up to 200° F, the 140° temperature is still within limits. Now, attach a heat sink and continue until the heat-sinked SCR exceeds 140° F. These experiments will demonstrate at what point a heat-sink is required and the effectiveness of the heat-sink of a given design.

6

Building Photo Detectors, Photo Couplers and Optical Fibre Devices

PHOTOSENSITIVITY

Many substances change their electrical characteristics in the presence of light. Changes are not only produced by visible light, some are more pronounced in the presence of invisible light, both infrared and ultraviolet. The two main classes of photosensitivity are, photoconductive (changing resistance) and photovoltaic (generating electric current). The photoconductive devices in the form of photo-electric cells have been around for years and have many useful applications. However, the photovoltaic, particularly the silicon devices, seem destined to predominate in many important present and future applications.

The silicon PN junction has been discussed above as a detector and rectifier. It is also photovoltaic and is available in a variety of forms. At a silicon PN junction, positive holes and negative electrons bind each other in what is known as a potential barrier. However, if light is directed at the junction, every photon of light liberates an electron and an electric current is generated.

THE FOUR MOST COMMON MODES

As optical detectors, silicon diodes and transistors are generally used in one of four modes. These modes are: back-biased diode; photovoltaic diode; phototransistor; and photo-Darlington as illustrated in Figures 6-1 through 6-4. Each mode has its own particular set of advantages and limitations and can be chosen for the best match of requirements and capabilities. The two most important characteristics are sensitivity and speed of response. Sensitivity, defined as current change per unit of light applied, increases in direct proportion to the area of the PN junction. However, a PN junction has capacitance which must be charged

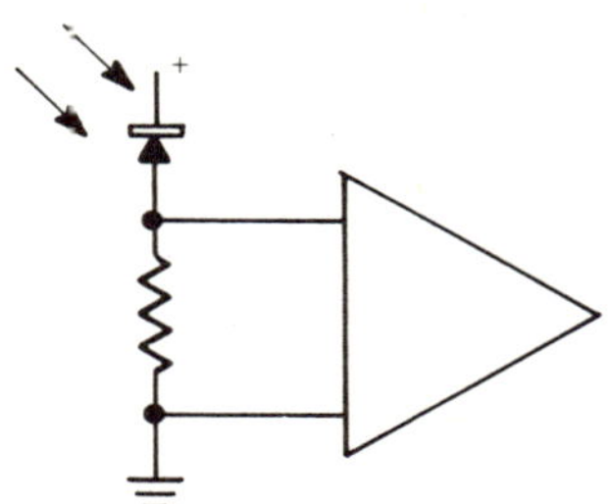

Figure 6-1—Back-Biased Diode

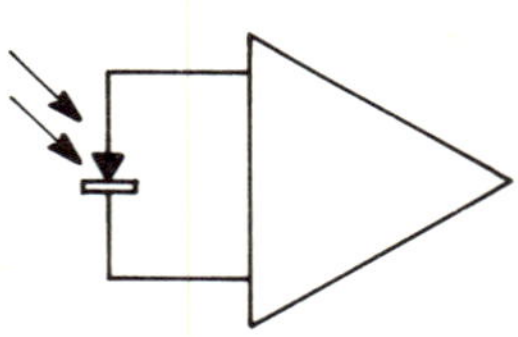

Figure 6-2—Photovoltaic Diode

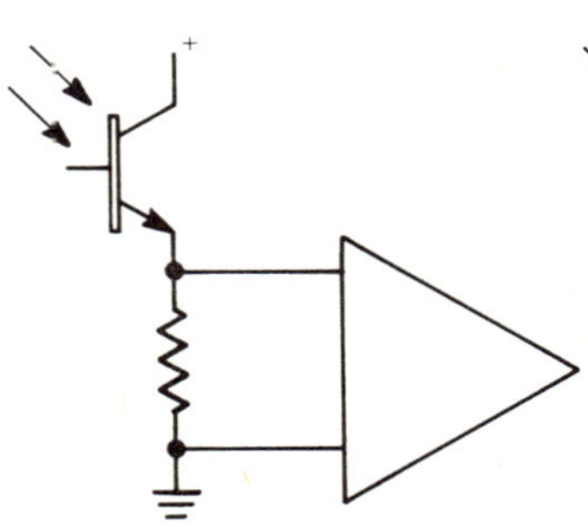

Figure 6-3—Phototransistor

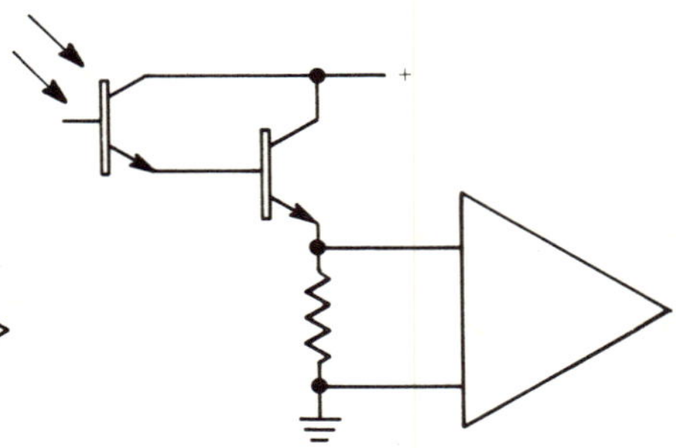

Figure 6-4—Photo-Darlington

and discharged when the light changes, and hence slows the response. Adding a transistor increases the capacitance, and adding a second transistor to form a Darlington increase the capacitance still further. However, back-biasing the junction, as shown in Figure 6-1, decreases the capacitance, making the back-biased photodiode the fastest response mode.

PROJECT: Build a Light Meter

The current generated by a silicon PN junction is linearly proportional to its cross-sectional area and the light intensity measured in milliwatts per square centimeter, or foot candles.

In order to make a light meter, one needs three things: a PN junction current source; a current meter; and a standard for calibration.

The silicon solar cell is an ideal PN junction for the purpose. One that has an area of 2 or 3 square centimeters is suggested. This will yield a short circuit current of 30 to 50 milliamperes maximum in bright sunlight. Figure 6-5 is a circuit with constants for a light meter having ranges from 0-0.2 mw/cm ^{2} to 200 mw/cm^2, which is sufficient for most experimental purposes. The meter and a four-position switch were mounted on a bakelite box along with the solar cell. The meter has a full-scale sensitivity of 50 microamperes and a resistance of 1740 ohms. Shunts are prepared by selecting, paralleling and series-connecting resistors to provide the required values as shown. A linear scale in mw/cm^2 may be drawn and placed on the face of the meter. Figure 6-6 shows the completed light meter.

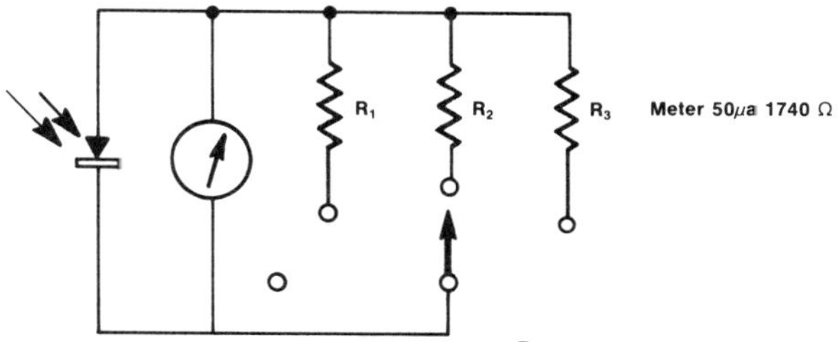

Range mw/cm²	Shunt	Factor
0-0.2	∞	1X
0-2	193.3Ω	X10
0-20	17.76Ω	X100
0-2000	1.74Ω	X1000

Figure 6-5—Circuit for Light Meter

Calibration: It is disarmingly easy to calibrate the light meter. The light from the sun at the earth's surface on a clear day is very close to 100 mw/cm^2, providing a very convenient source of known light intensity. If a solar cell is chosen (about 2 sq. cm.) which provides more than half-scale deflection on the least sensitive scale, it can be chipped down to a point where it provides the correct reading on the meter scale.

If one prefers to use the foot-candle convention, multiply all scale values by 20. Alternately provide two scales; one in milliwatts per square centimeter and the other in foot candles. However, the mw/cm^2 seems to be the more generally used standard.

The law which states that the current from a photovoltaic PN junction is proportional to light intensity assumes that the current is measured by a zero resistance meter. The light meter described above has resistance in the current meter. A true zero resistance meter is easily provided by means of an operational amplifier. However, the resistance of the silicon cell was found to be substantially higher than the meter resistance on all ranges, so that it introduced little error.

MEASURING PHOTOSENSITIVITY

Experimentation can lose much of its meaning if you do not know the characteristics of the components with which you are working. The two most meaningful characteristics of photodiodes and phototransistors are sensitivity and speed of response. The four basic photo-device/amplifier combinations are shown in Figure 6-1 through 6-4. Direct current sensitivity of each device is measured under stated conditions of incident light, device bias and load resistance and amplifier gain. The amplifier output is measured by means of a voltmeter, comprised of a current meter and resistor in series.

A known light source can be provided by means of the simple set-up shown in Figure 6-7. Using a 100-watt lamp, the shield when the lamp is placed about 3 feet away will expose 1 mw/cm^2 light intensity at the aperture. The calibration can be made with the light meter described above. Now, to use the device. In order to measure the sensitivity of a photo-device hold it in front of the 1-inch hole. The device will be exposed to a known standard of illumination, 1 mw/cm^2 (or 20 ft candles).

Figure 6-6—Light Meter Completed

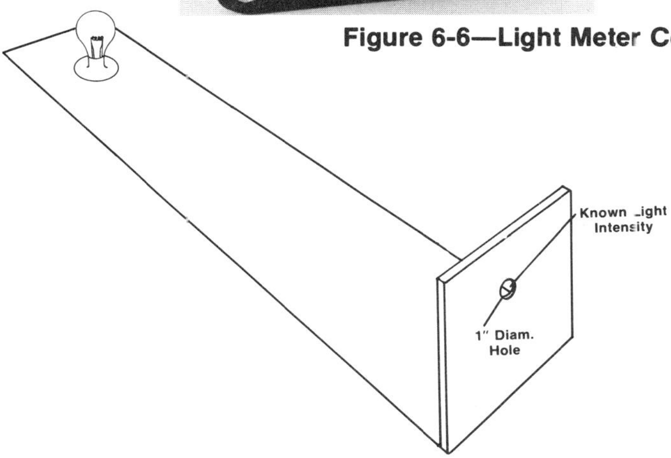

Figure 6-7—Set-Up for Providing Known Light Source

Experiments: With the standard illumination, measure the amplifier output voltage at a known amplifier gain. This output voltage E_o = IRA where I is the photo current, R is the load resistance in ohms and A is the amplifier gain. Expressing the sensitivity in current per milliwatt per square centimeter: $I = E_o/RA$. It can be expected that the phototransistors will be of an order of magnitude more sensitive than the photodiodes, and the photo-Darlington transistors are still another order more sensitive than the phototransistors.

MEASURING SPEED OF RESPONSE OF PHOTODEVICES

Many applications of photosensitive devices require fast response to changes in light intensity. Figure 6-8 is a partly block, partly simplifed schematic circuit diagram of a set-up for testing response speed of a phototransistor. A squarewave generator drives a small power transistor, which in turn drives a light emitting diode. The LED light source was chosen because it is known to respond to very fast changes in excitation. In this way, the phototransistor will be the slower device enabling an independent measurement of the phototransistor response. Of course the amplifier must have a fast slewing rate response so as not to limit the phototransistor measurements.

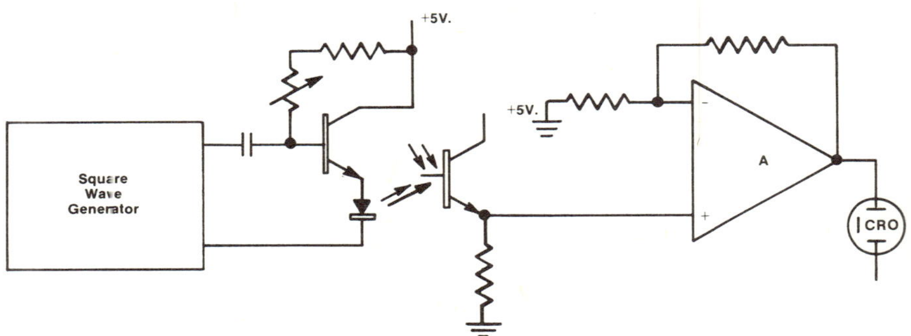

Figure 6-8—Set-Up for Testing Response Speed of Phototransistors

Experiments: Measure the response of the four devices of Figures 6-1, 6-2, 6-3 and 6-4. A convenient standard of rise time is the time required for the phototransistor collector current to rise from 10 percent to 90 percent of a peak value of 1 milliampere. When the output square wave on the CRO looks like the wave of Figure 6-9 as it just starts to decrease in amplitude, the rise time will be just about one-half the period of this wave. Thus, if the square wave frequency is 10KC, the rise time will be 1/10,000/2, or 0.00005 seconds which is the same as 50 microseconds.

Figure 6-9—Slewed Square Wave

The fastest photosensitive device is the photodiode. When the photodiode is back-biased as in Figure 6-1, the junction capacitance is decreased and the response becomes even faster.

Of course, the LED and photosensitive devices must be well shielded from ambient light generally rich in 60 cycle components.

PROJECT: Building an Optical Coupler

The basic circuit of the opto-coupler is shown in Figure 6-10. The LED and photo transistor should be enclosed in a lightproof housing. The emitting end of the LED should be facing and close to the photosensitive area of the phototransistor. The output device may be a photosensitive SCR as in Figure 6-11, or a phototransistor (Figure 6-12) or a photo-Darlington.

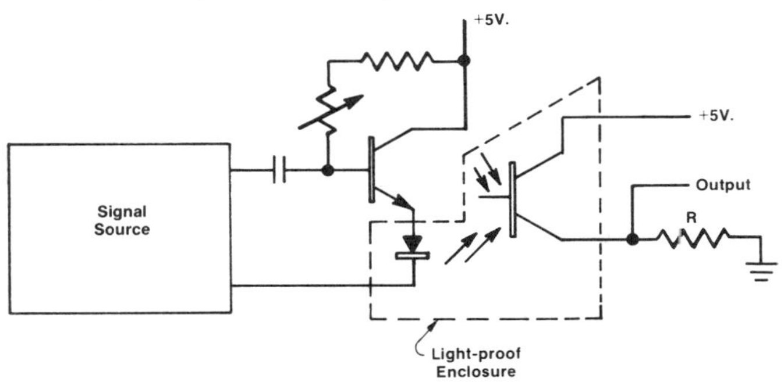

Figure 6-10—Circuit for an Optical Coupler

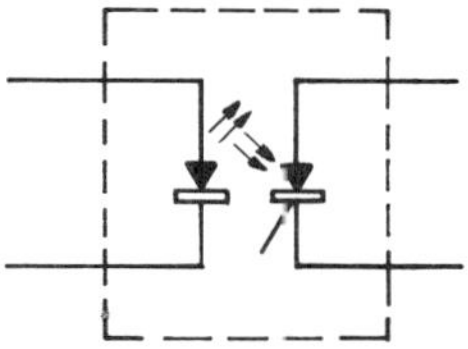

Figure 6-11—Photosensitive SCR

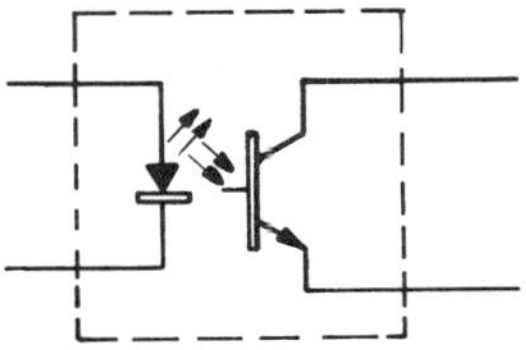

Figure 6-12—Phototransistor

Opto-couplers are used where conductive circuits must be interrupted. For example, a signal is received over a line which is not tied to a local ground. A conductive connection to such a line might result in introducing noise into the received signal. The solution is to use the signal to energize a LED and pick up the resulting light with a photosensitive device. Another typical application is in a system where there is a significant change in voltage level at some point. The opto-coupler bridges this voltage gap., Differences in level of thousands of volts can readily be accommodated by the use of such couplers. Furthermore, opto-couplers can handle either digital or analog signals.

LED'S IN OPTICAL TRANSMISSION

Measuring the Frequency Response of an LED

Light emitting diodes (LED's) are capable of being modulated at a high rate making them useful sources of light in analog communications systems. The question is, how can you be sure they will respond fast enough for your purposes. In other words, are they modulatable to, say, 20,000 hertz?

Experiment: Build a strobe light wheel as shown in Figure 6-13. A bakelite disc carrying a red LED is mounted on the shaft of a high-speed motor. One contact is made to the motor frame and the other to an insulated brush contact. The motor is rotated while audio signals are used to drive the LED through a single power transistor. The motor speed in the case of my experiment was 7000 r.p.m.* or 116.7 r.p.s.* The LED on a ⅞ inch radius or a circumference of 5.5 inches or 642 inches per second. Modulating the LED at 20,000 hertz produces a pattern 642/20,000 or 0.032 inches per hertz. This is readily observed as striations in the circular band of light produced by the rotating LED, proving a modulating capability of at least 20,000 hertz. Actually, observations were carried to 90,000 cycles, the final limitation being what could be observed even with a simple microscope.

Now, further experiments can be carried out, because you know first hand that the LED light source will not be the limiting factor as far as frequency response is concerned.

*r.p.s. and r.p.m. mean revolutions per second and revolutions per minute respectively.

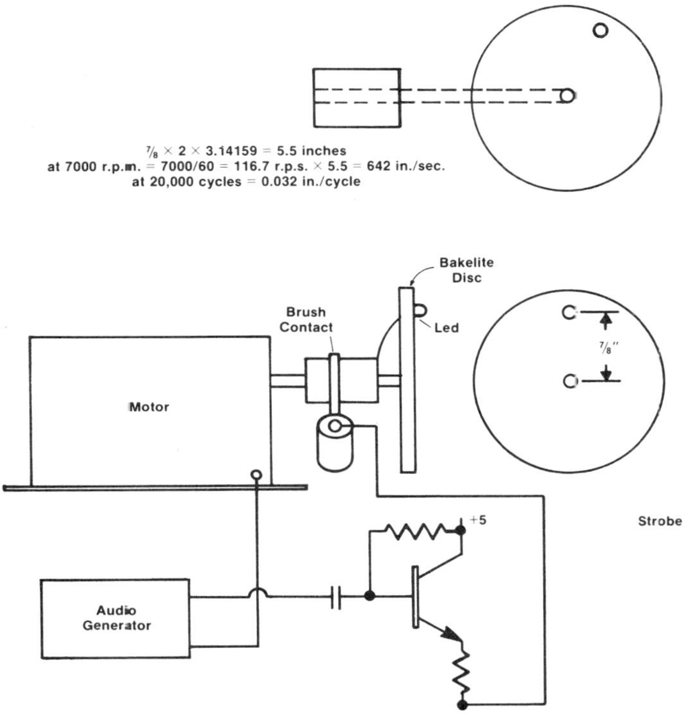

Figure 6-13—Strobe Light Wheel

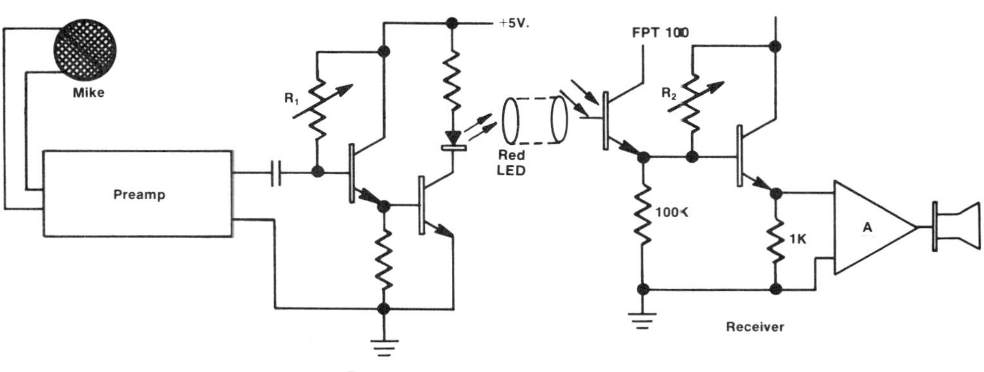

Figure 6-14—Set-Up for Optical Receiver/Transmitter

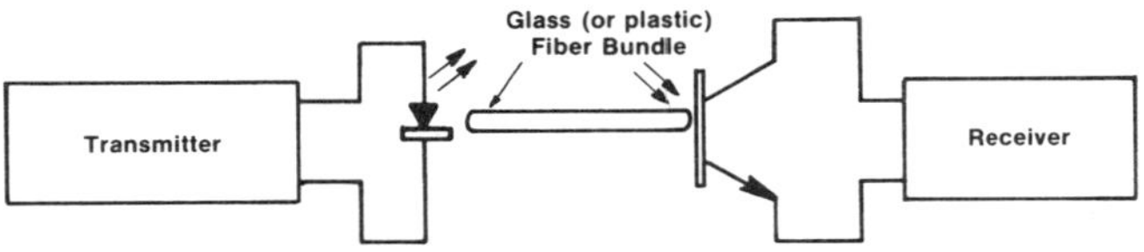

Figure 6-15—Glass or Fiber Couple Between LED and Phototransistor

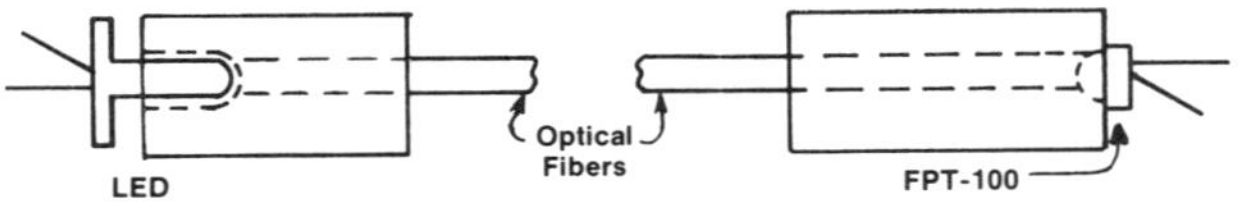

Figure 6-16—Plastic Coupling Shields

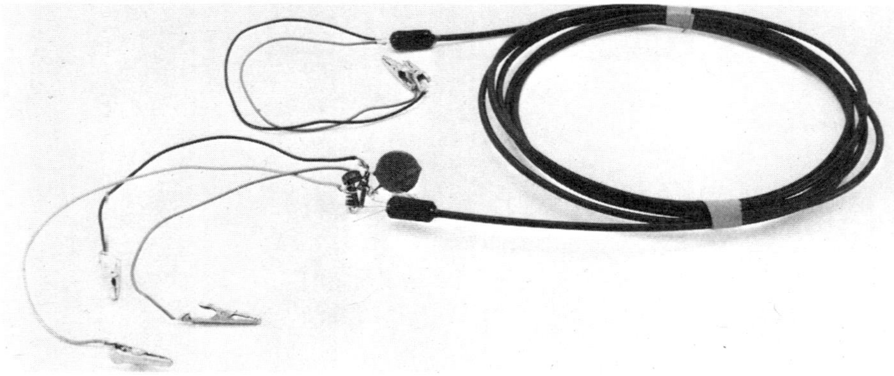

Figure 6-17—Optical Fiber Transmitter/Receiver

Talking over a Light Beam

Having determined that an LED can be modulated at audio frequencies and that a phototransistor can detect and amplify signals from an LED, the next step is to provide an optical path for the light. A simple path would be through the air, using two lenses as illustrated in Figure 6-14.

Experiment: Set up the optical transmitter and receiver as shown in Figure 6-14. While a red LED may not be as efficient as an infrared one, it has the advantage of visibility. First, get the LED to flicker when talking into the microphone showing voice modulation. Most linear operation is attained when R_1 is set to make the LED glow at half maximum instensity with no modulation.

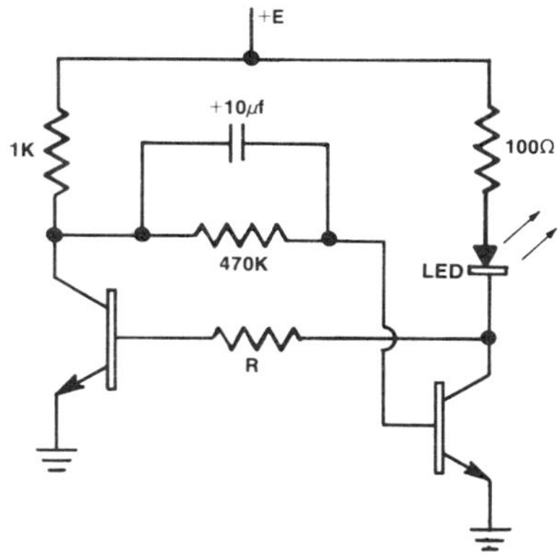

Battery	R	Pulse Rate
3V.		1 per second
4.5V.		1 per second
6V.		1 per second
9V.		1 per second

Figure 6-18—Circuit for Blinking Light Indicator

Once the transmitter is seen to be working, place the phototransistor in line and close to the LED without the lenses. With an oscilloscope across the 1K resistor, adjust R_2 for a symmetrical signal when talking into the microphone. Connect a loud speaker or tape recorder to the amplifier output and test for good speech reproduction through the system. To align the LED and phototransistor and shield from ambient light a short length of opaque plastic tubing can be used.

Now, when you have successfully operated the system, separate the LED and phototransistor a few inches and couple the two short focal length lenses as shown. This may not work very well for two reasons. First, the efficiency of light transmission may become too low to be practical and, second, 60 hertz light sources generally present may cause a masking hum signal. For such a system to be practical a much more powerful light source is required as well as a sophisticated optical system. Furthermore, you are limited to line of sight in this case.

Talking over Optical Fibres

A practical way to efficiently guide and transmit light is by means of optical fibres. A bundle of fibers can transmit light and be shielded from environmental light interference. A fine filament of glass transmits light along its length and around corners by means of very efficient internal reflections from the walls of the fiber. To provide increased light flow, a large number of glass fibers are bundled together. Typical fibers are 5 to 10 one-thousandths of an inch in diameter. To make a practical optical light guide many (100-200) fibers are bundled together inside an opaque plastic jacket.

It has long been found that certain plastics, too, transmit light in the same way as glass. Generally, plastic fibers are much cheaper than glass and more readily available.

Experiment: Couple the LED to the phototransistor with a bundle of plastic (or glass) fibers as shown in Figure 6-15. In order to align the LED and phototransistor and complete the ambient light shielding, plastic couplers as shown in Figure 6-16 should be used. Two concentric holes are drilled in pieces of opaque plastic rod material. One hole is sized to a snug fit with the plastic fiber and the other is sized to the LED or phototransistor. The plastic fiber should be pushed into contact with the devices at each end, and both should be cemented in place. Five-minute Epoxy is a convenient cement.

Now, talk over the system including the fiber optical path. For this experiment, a red LED, an FPT-100 or equivalent phototransistor and 10 or so feet of plastic fiber bundle of say, 0.0004 sq. in. cross-section is suggested. Substantially greater lengths are possible, but it should be remembered that optical fibers exhibit substantial amounts of attenuation per foot of length (unlike communicating electrically over wire lines). Loss figures are typically 20 percent per foot, plus 10 percent in the couplings at each end. This optical transmission system is shown in Figure 6-17.

BLINKING LIGHT ON INDICATOR FOR BATTERY-OPERATED EQUIPMENT

Many battery-operated devices have a common problem; battery life is severely limited. A blinking red light (LED) on such equipment is a very practical reminder to turn it off.

Experiment: Build the circuit shown in Figure 6-18 on a small piece of perforated board. Determine R for various battery voltages, E, to cause the LED to pulse approximately once per second. This device can be built on a very small board and tucked away in almost any piece of battery operated equipment. The LED, of course, should be mounted where it can be seen from any direction. An LED having a diffused glow provides the best visibility.

7

Projects and Experiments with Solar Cells

SOLAR CELL THEORY

The silicon solar cell or battery is a PN junction formed close to the front surface of a plate of nearly pure silicon. Silicon is one of our most common elements, occurring in the form of silicon dioxide, which is also known as ordinary sand. Pure silicon is used with a trace of an element from the fourth column of the periodic tables, such as arsenic, added as an n-type semiconductor containing free electrons. Each added arsenic atom frees one electron (unit negative charge) and leaves behind the arsenic atom with one unit positive charge bound into the crystal structure. Thus, n-type silicon consists of silicon to which are added equal numbers of free electrons and bound positive charges, so that there is no net charge.

Adding minute amounts of an element from the third column such as boron, aluminum or gallium, the silicon is made p-type and electric current is carried by holes. Analogous to the n-type, the p-type silicon consists of silicon having equal numbers of free positive holes and bound negative atomic charges, so that there is no net charge.

The interface between a p- and n-type silicon layer is called a PN junction. The free holes and free electrons try to intermix like gases.

However, the holes entering the n-type material disappear, leaving negatively charged gallium atoms, and the electrons which enter the p-type material disappear, leaving positively charged arsenic atoms. These charges form an electrical barrier or field which prevents the remaining holes and electrons from crossing the junction and mixing.

The solar cell body is n-type material carrying a p-type skin of about 0.0001 inch thickness. Photons of light from the sun produce hole-electron pairs near the junction. The fields described above force the holes to the p-side and the electrons to the n-side. This displacement of free charges results in a voltage difference between the two regions, the p-region being positive (plus) and the n-region negative (minus). When a circuit is completed between the p and n regions, an electron current flows, providing useful electrical power.

PRACTICAL CHARACTERISTICS OF SOLAR CELLS

Solar cells generate electric current which is proportional to the received light and the exposed area of the cell. Thus, for a given cell area, current generated by a solar cell is proportional to the received light. The voltage generated increases rapidly as the light intensity is increased and becomes asymptotic to approximately 0.5 volt. The open-circuit voltage decreases about 0.002 volt/°C as the

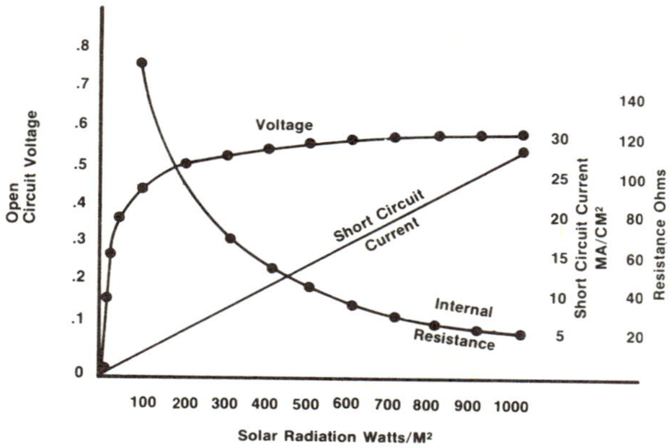

Figure 7-1—Characteristic Output of Solar Cells

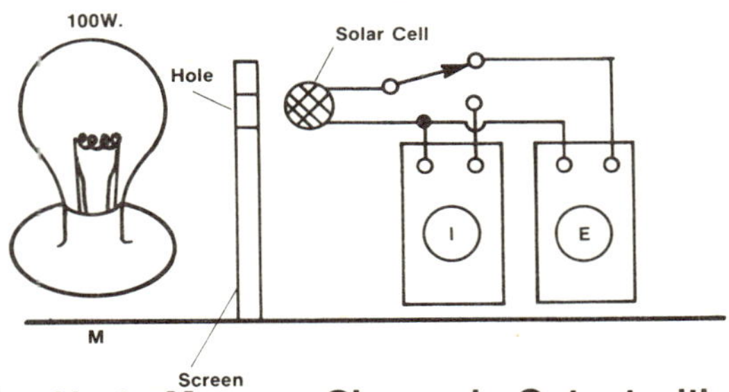

Figure 7-2—Set-Up to Measure Change in Output with Change in Light Intensity

temperature of the cell is increased. The internal resistance of the cell also changes with temperature.

Experiment: Measure the characteristics of solar cells with changes in light intensity. A set-up as shown in Figure 7-2 is suggested. A 100-watt incandescent lamp provides a convenient light source, since its intensity equals bright sunlight at a distance of 1 inch or so. Measure the light intensity with a light meter in milliwatts per square centimeter (mw/cm^2). Record E and I. Calculate the apparent internal resistance of the cell (E/I). Plot I, E and R versus light intensity. The result should be similar to the curves of Figure 7-1, but on a scale depending mainly on the area of the solar cell used in the experiment.

POWER FROM SOLAR CELLS

While the current generated by a solar cell is interesting, it is not directly useful as a power source. In the first place, current is measured into a short circuit which is not a useful load. Power delivered is the voltage across the load times the current through the load. The maximum power output from a solar cell depends on its generated voltage and internal resistance. The maximum efficiency provided by a solar cell is when the load resistance equals the internal resistance of the cell. The maximum power output is provided to the load having twice the resistance of the solar cell.

Experiment: Measure the power which can be derived from a solar cell using a set-up as shown in Figure 7-3. Record I and E for various values of load resistance R and plot the results. Select other light intensities and repeat. Calculate and plot power output for various light values. Some typical results are plotted in Figure 7-4. What conclusions can you draw from these results?

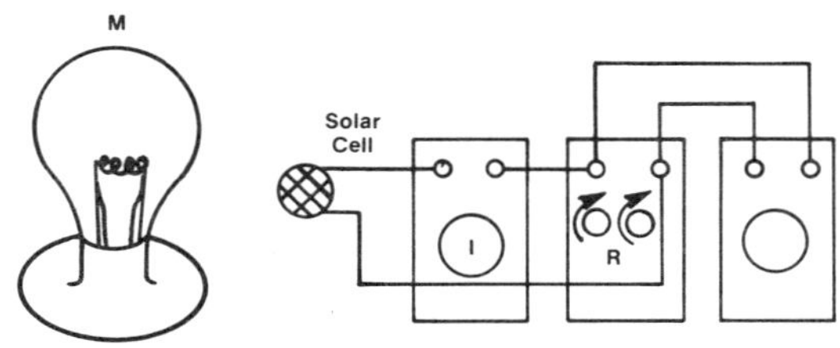

Figure 7-3—Measuring Power Derived from Solar Cells

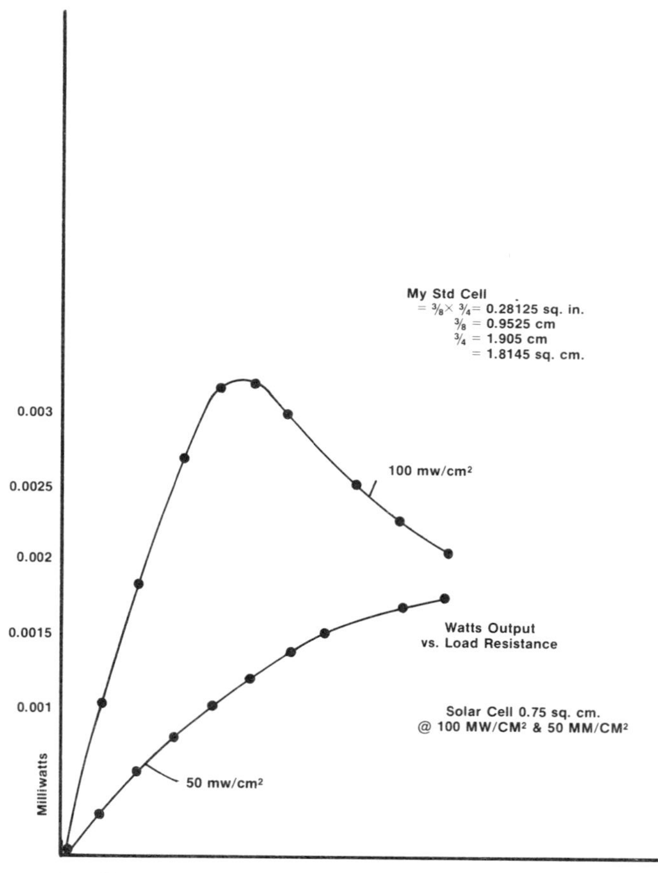

Figure 7-4—Plotted Results of Power Derived from Solar Cells

HOW TEMPERATURE AFFECTS SOLAR CELLS

As stated above, the open circuit voltage of a solar cell decreases about 0.002v/°C as the temperature is increased. One of the practical problems encountered in the use of solar cells is this decrease in voltage with temperature. The problem is particularly important in solar cell arrays employing light-concentrating means such as lenses or reflectors.

Experiment: In order to measure the effect of heat on a solar cell, means must be provided to supply a steady illumination to the cell while heating the cell to various predetermined temperatures. One way to do these things is illustrated in Figure 7-5. The cell is immersed in high-temperature cooking oil which can be heated to something on the order of 200°C. Heat is applied in increments and temperature, E and I are observed simultaneously, recorded and the results are plotted. A variation is to also measure the power output to various load resistances in order to find the effect of heat on the power output.

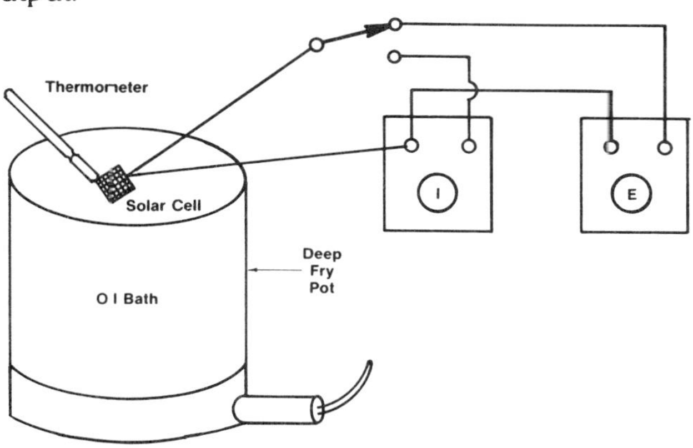

Figure 7-5—Measuring the Effect of Heat on Solar Cells

SOLAR CELLS AS POWER OR CURRENT SOURCES

Solar cells are commonly used in two different modes; as sources of current to charge batteries or as sources of power for direct use. Each mode has its own advantages and limitations. Since power sources are generally required at any or all times, solar cell power systems must operate where the sun shines steadily or at least when power is required. On the other hand, batteries may be charged in areas where sunshine is intermittent. A hybrid system is also possible where power is used to create stored power as by pumping water into a storage reservoir for generating hydro-electric power or for irrigation.

Charging Batteries Using Solar Cells

Charging batteries by means of solar cells is simple and practical in many cases. Equipment in remote areas, on camping trips or out in a boat are prime applications for solar cell battery charging. Solar cells of good construction generate just about 0.5 volts in bright sunlight. As can be seen from Figure 7-1, this voltage drops slowly to about 0.4 volt at 20mv/cm^2, after which it falls rapidly. If it is assumed that active charging will continue to about 0.4 volt at, say, 20mv/cm^2, the number of solar cells required for a practical charger can be calculated. A basic charging circuit is shown in Figure 7-6. Six cells at 0.4 volt provides 2.4 volts, less the protective diode drop of 0.5 volt leaves a net charging voltage of 1.9 volts sufficient for 1.2 or 1.5 volt batteries. The diode is necessary to prevent the battery from discharging through the solar cells when the generated voltage drops below that required to provide forward charging current. It should be realized that although charging will be possible down to light values of the order of 20mw/cm^2, the charging current will decrease in proportion to the decrease in light

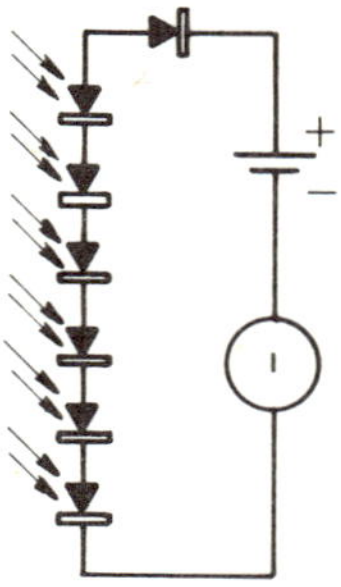

Figure 7-6—Basic Charging Circuit

intensity. Thus, if the charging current is 100 ma at 100mw/cm^2 it will only be 20 ma at 20mw/cm^2 and will fall rapidly to zero for lower values.

Experiments: Set up the charging circuit as shown in Figure 7-6. Observe the charging current under various conditions of sunlight, haze, light clouds and so forth. Observe the charging current when using a light bulb source and vary the distance between cells and lamp. How does the intensity of light required to charge a battery compare with that required for reading and so on. Any 1.2 or 1.5 volt cell which is rechargeable is suitable for these experiments. The 1.2 volt Ni Cad is generally available and well suited to the purpose. If a lead acid 2.0 volt cell is used, a seventh solar cell should be added to the string of cells.

If the charger is placed where it receives strong sunlight all day, it is a simple matter to evaluate the charge by multiplying the charging current by the number of daylight hours. Say, 100 ma times 10 hours, which equals 1 ampere hour per day. However, if the charger is located where it does not receive maximum sunlight all day or the day is partly cloudy, then the charge is harder to evaluate.

Actually, the charging is a type of integration process, and this must be taken into consideration. One way to obtain an evaluation of the integrated charge is to charge the battery and then to check the accumulated charge. Figure 7-7 is a circuit for evaluating the charge. The battery after charging is discharged at a predetermined rate, and the time taken to discharge is measured. Resistor R_1 is varied to set a predetermined discharge rate, say, 50 milliamperes and R_2 is set so that the relay opens at a predetermined discharged voltage point. This discharge point might be 1 volt for a NiCad battery or 1.6 volts for a lead acid

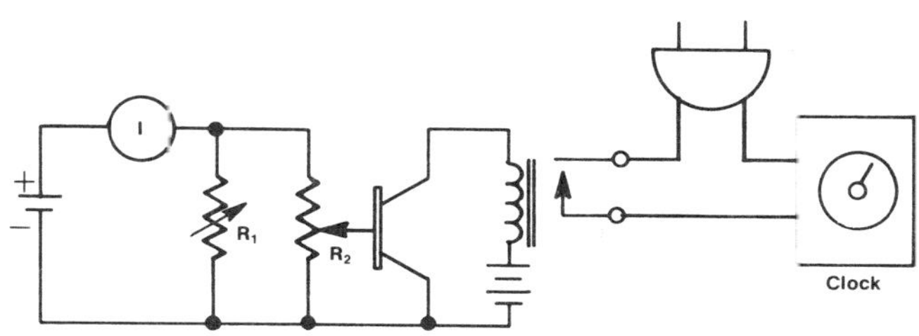

Figure 7-7—Circuit for Evaluating the Charge

battery. The clock is initially set at 12 o'clock, and the time registered when the battery is discharged is the number of discharge hours. The number of discharge hours multiplied by the discharge rate is the ampere hours of accumulated charge. Such tests can be run for, say, a day, two days or a week, depending on the sun conditions.

It is important to note that the charging current will be determined by the lowest current-generating cell in the string. Hence, for efficient predictable charging, all cells should be of the same size and current-generating characteristics. If one or more cells receive less light than the others, as by a shadow, the output current will be similarly limited by the cell supplying the least current. An open cell will, of course, reduce the output current to zero, while a shorted cell will merely reduce the output voltage by an amount equal to the terminal voltage of that particular cell.

Direct Power from Solar Cells

Most of the characteristics and conditions for solar cell generation of electrical current set forth above hold equally for the direct power mode of solar cell use. The sun has to be shining in order to generate useful power. The most obvious application of direct solar cell power is in space vehicles and satellites which operate outside the earth's atmosphere and even outside the shadow of the earth.

Experiment: One of the experiments which can be performed with direct solar cell power is the operation of portable electronic equipment shown in Figure 7-8. A conservative calculation can be based on a cell voltage of 0.4 volt which, with a protective diode drop of 0.5 volt yields the equation N = E+0.5/0.4 where N is the

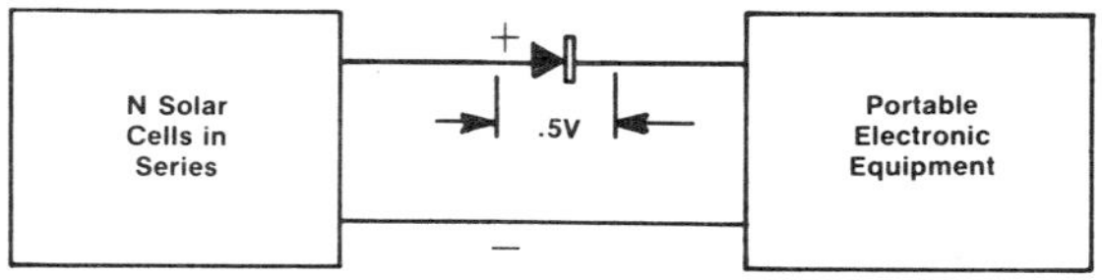

$$\text{Number of Cells} = \frac{\text{Output Voltage} + 0.5}{0.4}$$

Figure 7-8—Operation of Portable Equipment from Solar Cells

number of solar cells required for a voltage to the equipment to be powered of E volts. Applying this formula, commonly used voltages can be generated by the number of cells shown in the table below:

Equipment Voltage	Number of Solar Cells
3.1	9
4.3	12
6.2	17
9.1	24

The portable equipment that can be directly powered from solar cells will depend on the voltage and current required to power the equipment, the size of the solar cells and the presence of direct sunlight.

PROJECT: Build a Solar Cell Battery for a Portable Radio
Figure 7-9 is a photograph of a 12-cell solar panel providing direct power to a portable transistor radio designed to operate on a 6-volt battery. The individual cells of the panel are approximately 1×2 cm and deliver 25 ma at 0.5 volt in bright sunlight. The radio draws 10 ma at 6 volts on low volume, peaking to 25 ma at high

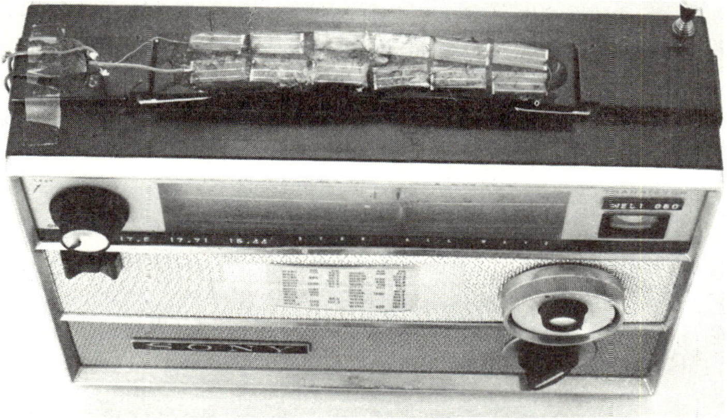

Figure 7-9—Solar Powered Portable Radio

volume. The operation with the solar cell panel is very satisfactory as long as the sun is shining.

In order to provide more current, either larger cells or parallel panels of cells must be used. Figure 7-10 is a circuit for observing the operation of solar cells in parallel. Try covering one or two cells and observe the effect on the circulating current, I, and the output voltage, E. Connect the two strings in parallel and connect them across the current meter. Observe the output current as one or two cells are covered. What have you learned about the characteristics of parallel-connected solar panels?

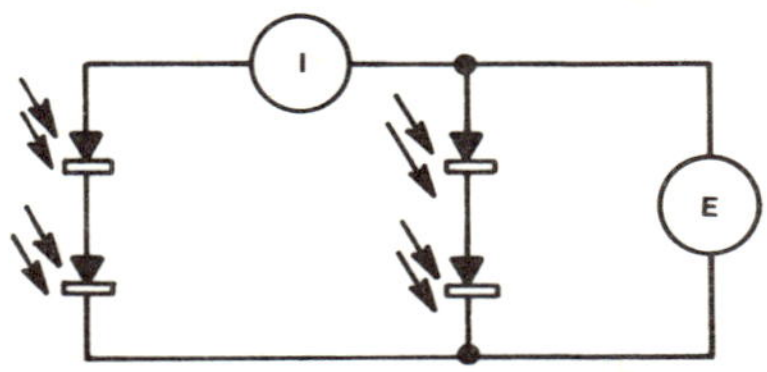

Figure 7-10—Circuit for Observing Solar Cells in Parallel

HIGH RELIABILITY SOLAR POWER

Solar cell battery chargers are used to provide power to operate equipment in remote areas. The average sunlight hours are used to compute the size of cell required. It is quite evident that reliability in such a system is of utmost importance.

A solar cell is a silicon diode. Light shining on the cell generates an electric current. The anode of the cell is positive and current flows as shown in Figure 7-11. However, suppose no light falls on one of the cells. It will be back-biased by the current from the other cell as shown in Figure 7-12 and no current or a very small current will flow. In other words, for a solar battery to operate normally, all cells in the series string must receive substantially the same amount of light. Thus, in a system, if only a single series string of solar cells are employed, the system may be shut down if just one cell fails to receive light or becomes inoperative.

An improvement can be effected by using two sets of cells connected in parallel as shown in Figure 7-13. In this case, the current is doubled at the expense of twice as many cells. The reliability is improved, but not completely, because if any one cell fails, the current is cut in half. If two cells in parallel fail, the voltage source becomes inoperative.

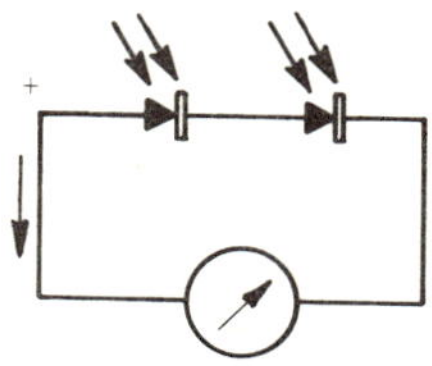

Figure 7-11—Current Flow in a Solar Cell

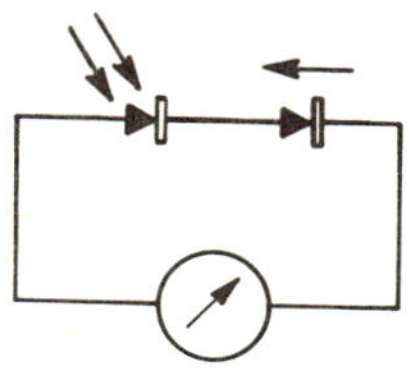

Figure 7-12—Current Flow with Only One Cell Working

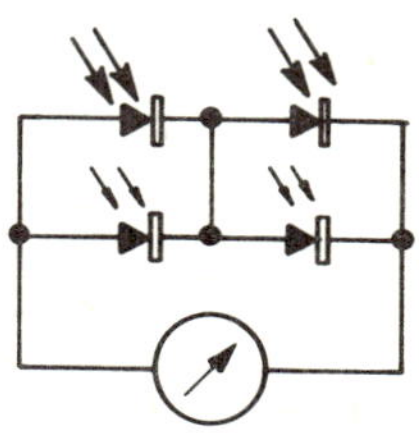

Figure 7-13—Two Sets of Solar Cells in Parallel

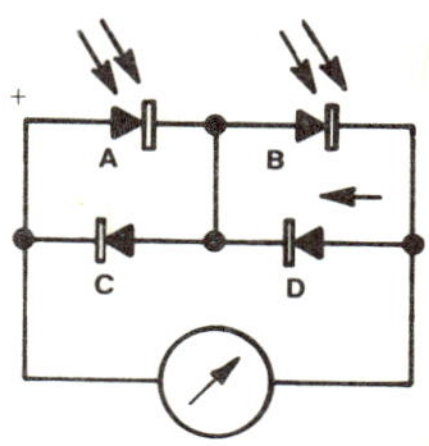

Figure 7-14—Circuit to Effect Drop in Voltage Instead of Current

Rather than a drop in current, there is a way to experience a drop in voltage instead when a cell is cut off. An arrangement to accomplish this is shown in Figure 7-14. Two solar cells, A and B, are shunted by two diodes, C and D, which are connected so that if one cell fails, the current is maintained by its shunting diode. For example, if solar cell B fails, the current from cell A will be carried by diode D. In order to maintain a given current in a battery charging system, for example, all that is required is the addition of enough series cells to maintain a given voltage in case of the failure of one or even several cells. This is a practical and highly reliable arrangement and much less expensive than doubling the number of cells. Furthermore, the current is maintained, which is the more important factor. For example, in a 30-cell battery charger, merely by adding three more cells and diodes, a system is provided in which any three cells can become inoperative without decreasing the charging current by more than 10 percent. With 60 cells arranged in two parallel strings and cross connected, if even one cell becomes inoperative, the current is reduced 50 percent.

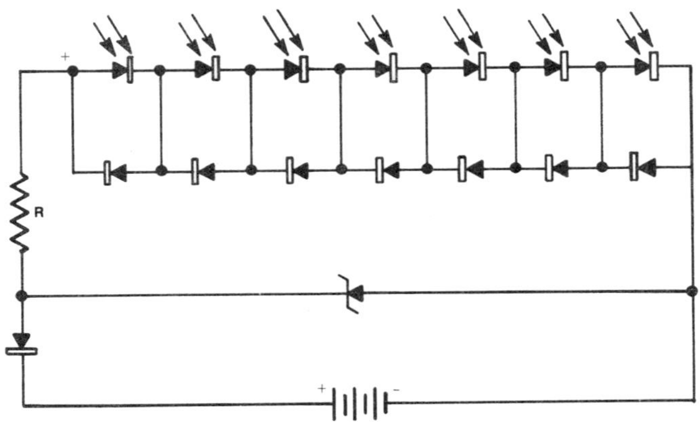

Figure 7-15—Circuit of Battery Charger Employing Solar Cells and Diode Shunts

Figure 7-15 is a schematic circuit diagram of a battery charger employing solar cells and diode shunts. A series resistor, R, and a zener diode provide voltage and current regulation for a very precise battery charge. The added diode prevents discharge of the battery through the zener diode during conditions of low or no sunlight.

This solar cell/diode system has a very practical application in the case of solar cell-operated portable equipment, such as a portable radio. In this case, the radio is connected in place of the battery. Without the diodes, even a small shadow on one of the solar cells will cut off operation. This can be very disconcerting. With a few extra cells, the diodes and zener, small and even moderate shadows will not affect the operation of the radio. For greatest immunity to shadows, the solar cells should be strung out in as long a line as convenient. Figure 7-9 shows a portable radio with a solar cell battery along the carrying handle. Incidentally, this radio operates very well even from the light of a table lamp in a hotel or motel room.

8

Projects and Experiments with Field Effect Transistors

The field effect transistor (FET) was invented and developed at almost the same time as the bipolar transistor. Instead of bipolar conduction of electrons and holes across PN junctions, the FET is an electron flow or unipolar device in which conduction is controlled by an electrostatic field. (See Figure 8-1) In this way, and for other reasons, the FET comes much closer to being analogous to the vacuum tube than does the bipolar transistor.

The first attempt to control electron flow in a semiconductor was in a device called the field-sistor. In this device, an electrostatic field was created around a point contact PN junction in order to control current flow across the junction. However, the field-sistor was very critical mechanically and generally impractical. Hence, it never got beyond the experimental state. The idea did not die, however, and many very useful field effect devices were eventually developed.

UNDERSTANDING THE FET

The basic concept of the FET is quite simple. A bar of silicon called a channel carries current, which is controlled by an electrical field, along the middle of the bar, called a gate (See Figure 8-1). One end of the channel is called the drain (analogous to the plate of the vacuum tube) and the other end is called the source (analogous to the cathode of a vacuum tube). Actually, the channel is deposited or grown on a semi-conductor substrate and the gate is deposited or grown on top of the channel. The channel may be either N-type or P-type silicon with the gate P- or N-type respectively. While the gate and channel would appear to form a PN junction, control of the current between source and drain is electrostatic and the PN junction is reverse-biased. (In a bipolar transistor, the PN junction is forward-biased.) This back-bias operation results in very low drain current and, essentially, an electrostatic voltage control as found in a vacuum tube.

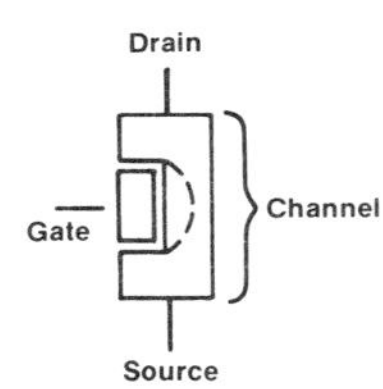

Figure 8-1—Diagram of Field Effect Transistor (FET)

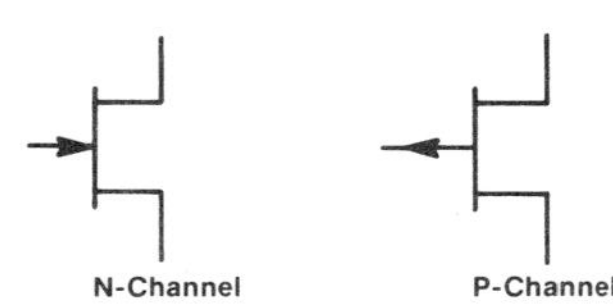

Figure 8-2—Symbols for N-Channel FET and P-Channel FET

Symbols for the two channel types, namely, N-channel and P-channel are shown in Figure 8-2. There are three designated variations of each type; enhancement, depletion and enhancement/depletion types. Enchancement means the FET passes little or no current at zero gate bias and that the gate voltage is to be increased in the direction of the drain voltage to cause controlled conduction. Depletion means that the FET conducts at zero gate voltage and is controlled by varying the gate voltage in the opposite direction to the polarity of the drain voltage. The enhancement/depletion type conducts at zero gate voltage and is controlled by gate voltages above and below zero.

Figure 8-3 is a diagram of a set-up for measuring the $I_D-E_G-I_G$ characteristics of a JFET (contact gate types are known as JFET's). Figure 8-4 is a plot of the drain current versus gate voltage for a sample type F0015 (HEP designation) JFET. It can be seen that this particular transistor operates in an enhancement/depletion mode and that it draws no gate current until the conduction voltage of the gate to channel diode is exceeded. This also illustrates why the JFET operates with a high gate impedance and can be considered as a voltage-operated device unlike bipolar transistors which exhibit base-to-emitter conduction making them current-operated devices.

Another characteristic of the JFET is its constant drain current versus drain voltage, much like a pentode vacuum tube. Figure 8-5 is a plot of data taken on the same N-channel JFET of Figure 8-4. The region below the knee is sometimes called the "triode" region and the area above that is the "pentode" region.

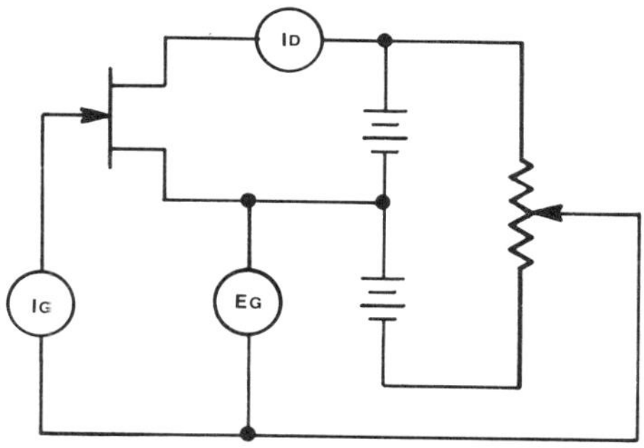

Figure 8-3—Set-Up for Measuring the I_D, E_G, and I_G of a JFET

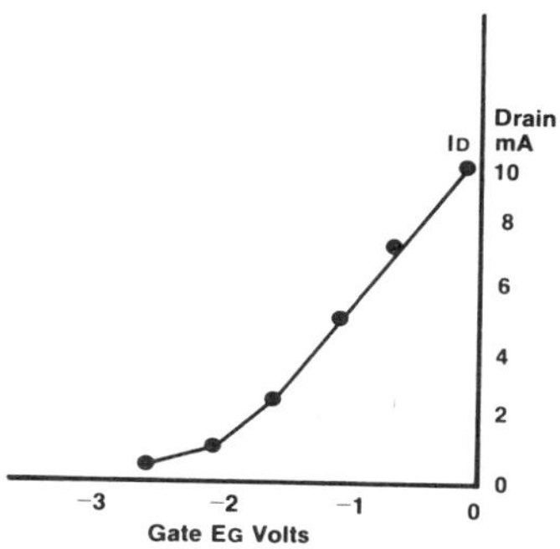

Figure 8-4—Drain Current Versus Gate Voltage for ECG220 MOSFET

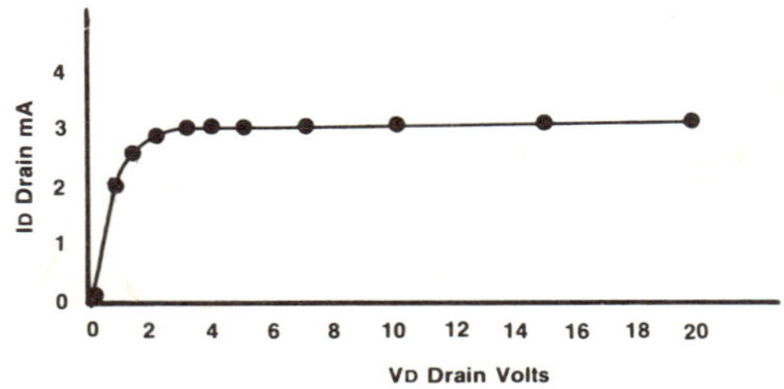

Figure 8-5—Current Drain Versus Voltage Drain

APPLICATIONS OF FETS

The unique characteristics of FETs makes them especially useful for voltage-controlled amplifiers, voltage-controlled resistances, cascade amplifiers, high-input impedance op amps, unneutralized rf amplifiers, high-input impedance followers, and high-input impedance drivers.

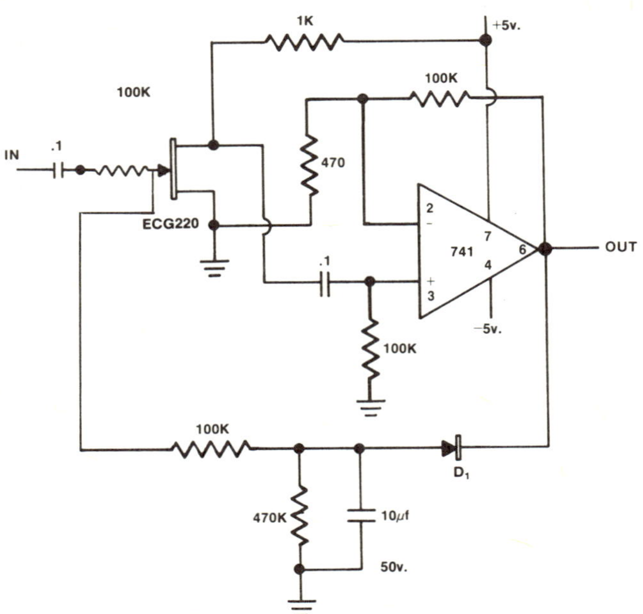

Figure 8-6—Circuit Diagram for Automatic Level Control Module

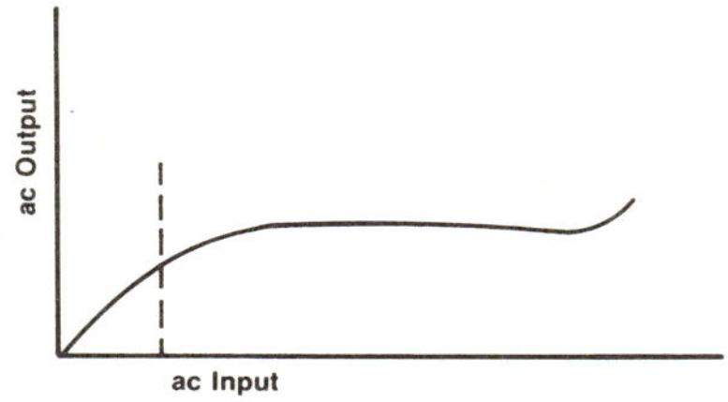

Figure 8-7—Controlled Output Versus ac Input

PROJECT: Building an Automatic Level Control Module

There are two ways in which an FET can be used in an automatic level control circuit. One is to use it as an element in an attenuator where it is controlled to act as a variable resistor. The other is to use it as a controlled gain amplifier stage. In either case it is important to use this gain control element at a low signal level point in the circuit to minimize distortion. Of course if the level is too low one will experience noise problems.

The gain controlled amplifier mode was chosen for this project. The circuit diagram is shown in Fig. 8-6. The gain controlled MOSFET is placed between the input line and the input to a 741 op amp which provides voltage gain and the amplified output. Since the 220 MOSFET is an N-channel depletion type of FET, its gain is maximum at zero bias. In order to control the gain, the output of the op amp is rectified by a diode D_1 producing a negative going voltage across the filter circuit consisting of the 10 mf capacitor and 470K resistor. This voltage is applied through the 100K decoupling resistor to the gate of the FET.

Initially the input signals are amplified with the full gain of the FET and the op amp in cascade. As the signal is increased, the diode D_1 will start to conduct building up a negative dc voltage across the capacitor. This negative voltage applied to the gate of the FET will start to reduce its gain. At some point the decrease in gain will almost balance out the increase in signal level and the output will become constant. This action is illustrated in Fig. 8-7. The dotted line indicates the point at which control starts.

MEASURING DISTORTION IN AN ALC CIRCUIT

While the JFET acts like a voltage-controlled resistance, it is a non-linear resistance. This means that it distorts the AC signal across its effective resistance. However, the smaller the signal across it, the smaller will be the distortion. The higher the gain of the op amp, the lower the signal across the FET and the smaller the distortion will be.

Figure 8-8 shows how the voltage across the control FET varies with input signal for two different values of op amp gain. Curve A is for an op amp gain of 100 and curve B for an op amp gain of 50 times. The effect on distortion of the amplitude of signal across the FET is shown in Figure 8-9. Curve A is for the higher gain and B for the lower. The improvement in distortion in A is due to the smaller voltage across the FET controlled resistance. If the dotted line at 1 percent distortion in Figure 8-9 is taken as the

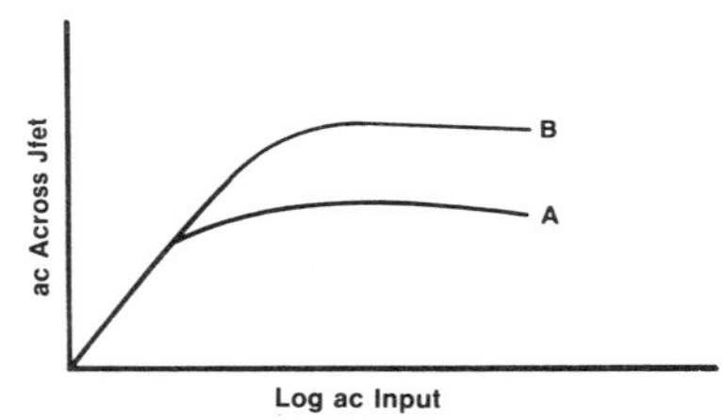

Figure 8-8—Log ac Input Versus ac Across JFET

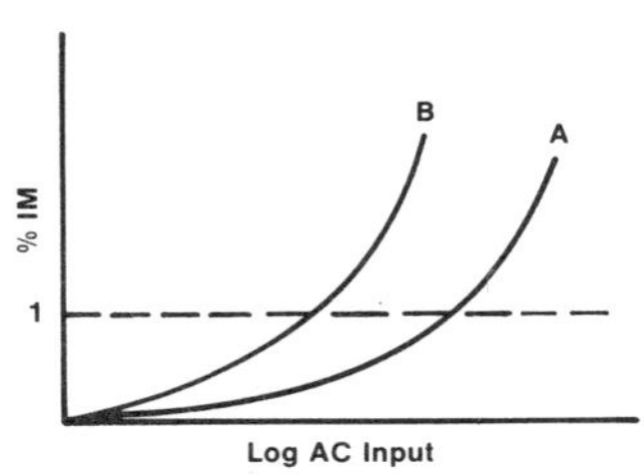

Figure 8-9—Effect on Distortion of Amplitude Signal

maximum tolerable distortion, the two different gain conditions result in 2+ decades of input signal for gain of 100 as against 1.5 decades for the lower gain.

EXPERIMENTING WITH TIME CONSTANTS IN AN ALC CIRCUIT

One of the most important characteristics of an ALC circuit which must be taken into consideration is its time constant. If an ALC circuit acted instantaneously at all times, it would have the effect of creating constant amplitude of all tones in music and speech, robbing them of all dynamic range and color The more useful characteristic is a fast attack and slow relaxation. This is

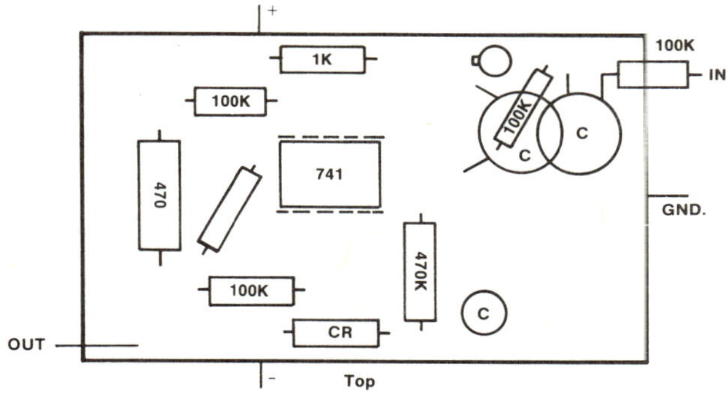

Figure 8-10—Top and Bottom Views of Automatic Level Control Module (See Also Next Page)

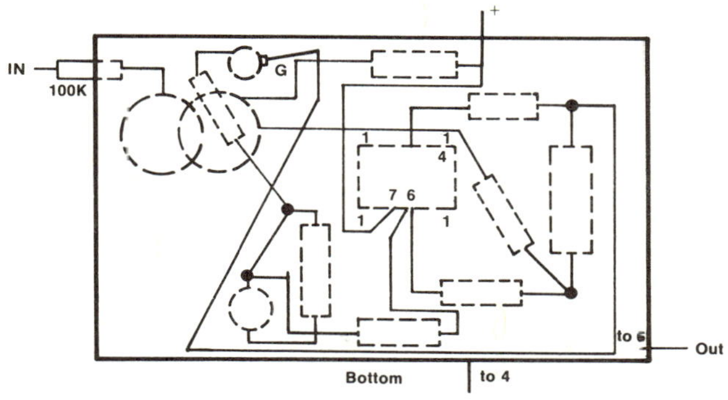

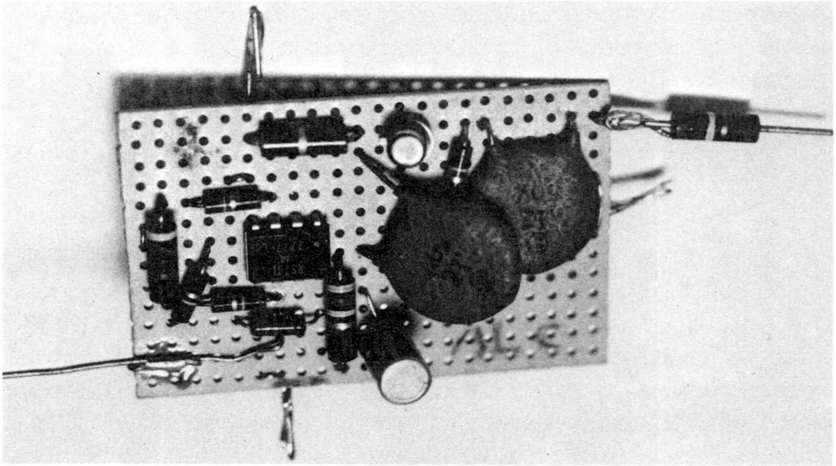

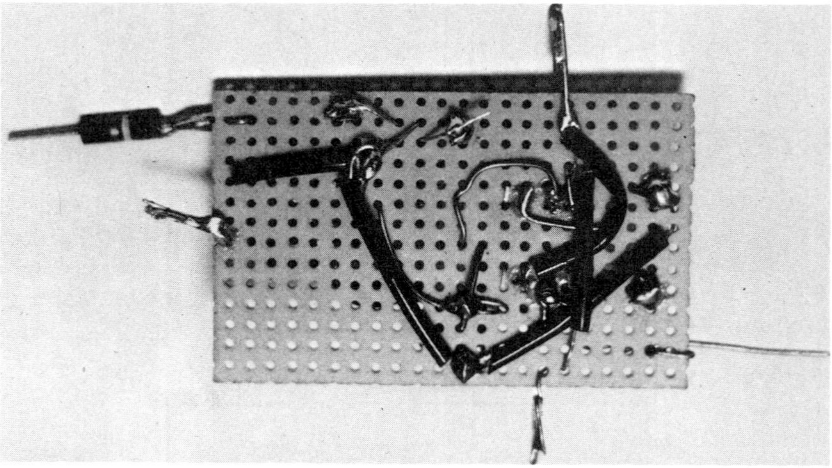

provided for in the circuit of Figure 8-6. The attack time, that is the time it takes for the automatic control to take control is the time required to charge the 10 microfarad capacitor. The charging series impedance is the internal impedance of the op amp output circuits, and the forward impedance of the rectifier. The relaxing time is the time constant of the capacitor and its shunting 470K resistor. Experiments with time constant can be made by providing an input signal from a microphone and preamplifier or some tape or record playback and an output through a power amplifier and loud speaker. With this set-up try changing the capacitor value to say 1 microfarad in one direction and 50 microfarads in the other. Listen for distortion due to a sudden increase in input level and the return to high gain due to decreasing the input level. In this way, one can evaluate the effect of time constant on music and speech.

The above measurements and experiments will indicate how ALC in portable cassette tape recorders causes distortion and restricts dynamic range, particularly in the recording of live music.

Figure 8-10 is a top and bottom photograph of an ALC circuit, made in accordance with the schematic of Figure 8-6 and on which data was taken for Figures 8-7, 8-8 and 8-9.

EXPERIMENTING WITH MOSFET TRANSISTORS

The naturally high-input resistance of the FET can be increased by several decades by insulating the gate creating what is known as the MOSFET. The usual way this is done is to deposit a very thin layer of silicon dioxide over the channel and on top of this to deposit a very thin layer of aluminum for the gate electrode. The symbol for a MOSFET is shown in Figure 8-11. The gate control here is also electrostatic, but since the gate is actually insulated, the

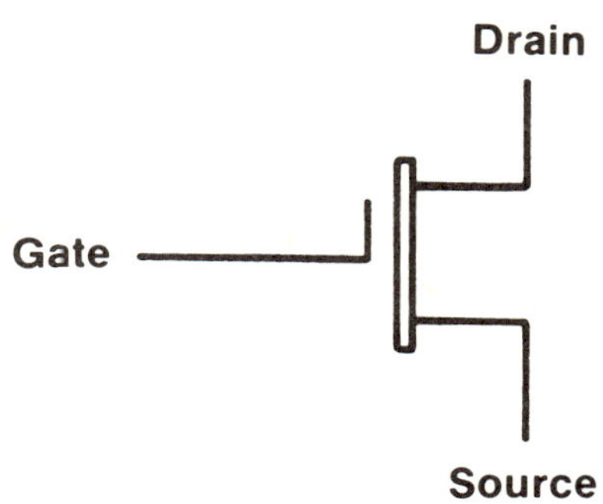

Figure 8-11—Symbol for a MOSFET

gate resistance is made very high, for example, 10^{-11} as against some 10^{-9} for a typical JFET.

The MOSFET, however, has one serious vulnerability, that is, the gate insulator is easily punctured, even by a minute current such as an electrostatic charge. Once the gate insulator is punctured, the MOSFET loses its high-gate resistance. Some of the precautions taken to guard against puncture include shipping with gate conductively grounded to the source, handling by the case without touching the gate lead and handling only in a special static-free environment. The best safe-guard, although at an increase in cost, is an internally integrated zener diode connected from gate to source as illustrated by the MOSFET of Figure 8-12.

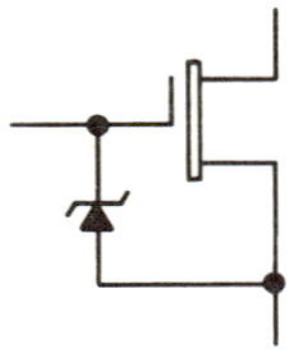

Figure 8-12—MOSFET with Integrated Zener Diode

Switching Applications of MOSFETs

The MOSFET makes a nearly ideal solid-state switch: It is voltage responsive, consumes almost no input power, so that a great many switches can be operated from a low output source; it is fast acting; and it exhibits a low ON resistance. These characteristics have led to the development and use of MOSFETS in digital-switching integrated circuits.

CONSTANT CURRENT USING A FIELD EFFECT TRANSISTOR

Figure 8-5 is a plot of data taken using an N-channel JFET and shows the very flat drain current versus drain voltage characteristic. This makes the JFET an effective and useful constant current device. Using a depletion mode JFET, this constant current can be set to a predetermined value by means of a variable source resistor as shown in Figure 8-13.

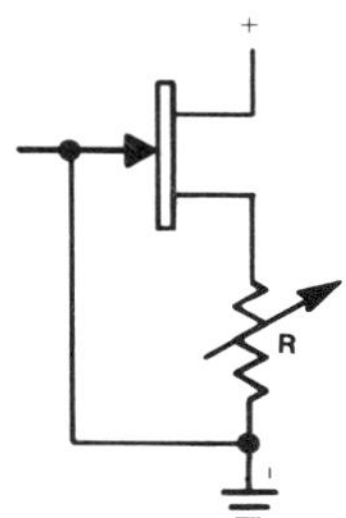

Figure 8-13—JFET Set to a Constant Current

UNDERSTANDING POWER FETS (VMOS)

Field effect transistors have certain properties which give them substantial advantages over bipolars. For emphasis, they are enumerated below:

1. Under gate control, power FETs exhibit positive temperature coefficients so that:

A. They readily operate in parallel without current-sharing resistors.

B. They do not tend to develop hot spots and so are not subject to secondary breakdown.

2. They have a fast response which reduces tendencies to develop transient intermodulation distortion.

3. They have very linear drain current versus gate voltage characteristics.

4. They are voltage responsive and therefore do not require current drivers, thus reducing component count and increasing reliability.

5. Low feedback capacitance makes possible unneutralized radio frequency amplifiers.

On the negative side, however, matched complementary devices may not be available, so that the more complicated quasi-complementary circuits must be used in audio power amplifiers;

maximum operating voltages may be limited; and they may be substantially more expensive watt for watt than equivalent bipolars.

USING THE TRANSISTOR TO MONITOR ITS OWN TEMPERATURE

The drain to source resistance of a VMOS transistor with its gate shorted to its source can be used to monitor its own temperature. A VMOS transistor is wired to the bulb of a high temperature thermometer, and the two are immersed in high temperature oil in a hot pot as shown in Figure 8-14. The resistance of the transistor drain to source is measured with a wide-range ohmmeter at temperature increments from room temperature to 150°C (302°F). The results are shown plotted in Figure 8-15. The temperature scale was linear and the transistor resistance was found to vary logarithmically from 10,000 megohms at 70°F to 470,000 ohms at 302°F.

The same procedure is used to calibrate a bipolar transistor (2N5295) resistance versus temperature. The resistance is measured from collector to emitter, reverse-bias, with base connected to emitter. The results are also plotted in Figure 8-15 with a dash line. In this case, the resistance varied from 25 megohms at room temperature to 10,000 ohms at 302°F.

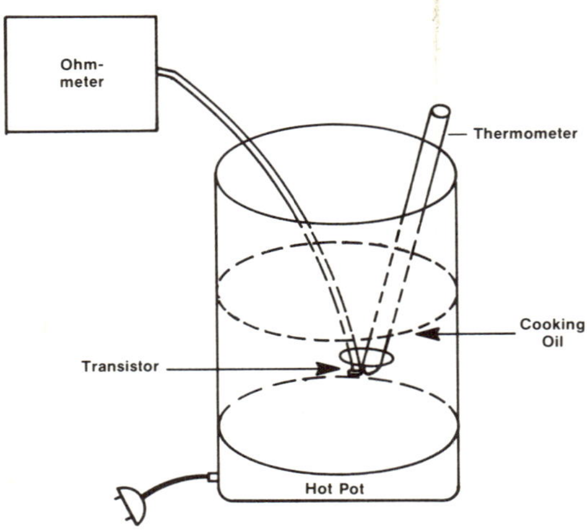

Figure 8-14—Immersing VMOS Transistor in Hot Oil to Test Variation According to Temperature

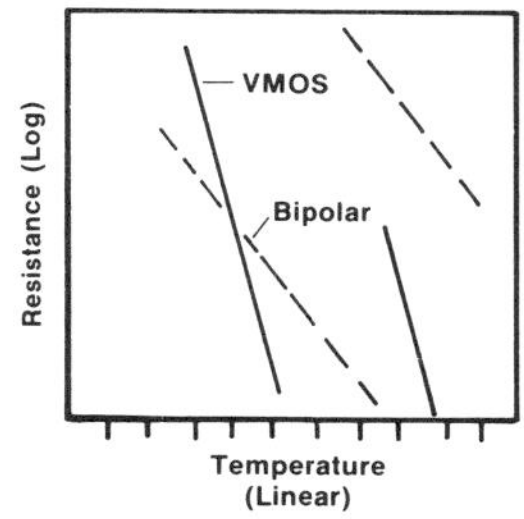

Figure 8-15—Plot of Functions of Temperature Versus Resistance

COMPARING TRANSISTORS WITH AND WITHOUT A HEATSINK

Since both the VN66AF and the bipolar 2N5295 are of similar construction known as TAB, they form the basis for some interesting comparisons. Two similar circuits for measuring temperature rise versus dissipation. Figure 8-16 for the VN66AF and Figure 8-17 for the 2N5295. Varying amounts of power could be applied by varying E or R. The temperature was measured by removing the power and switching in the ohmmeter using a double pole-double throw switch. The resistance readings had to be made very quickly on switching since heat flow from the tiny transistor junctions is very rapid. Power was applied at several levels and left on until the junction temperature had stabilized. A first run with each transistor was made at reduced power and no heatsink. The thermal

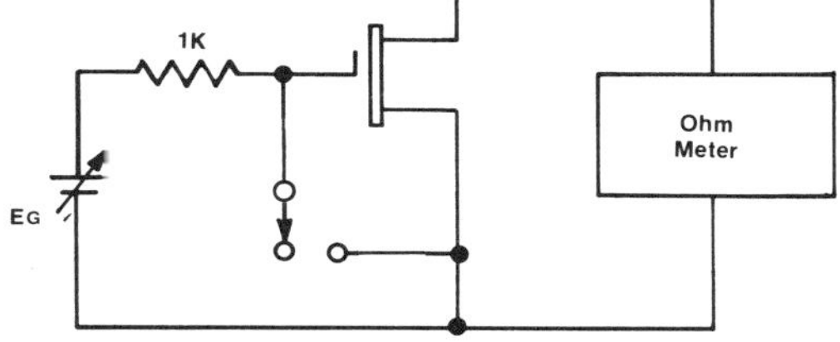

Figure 8-16—Circuit for Measuring Temperature Rise Versus Dissipation for VN66AF

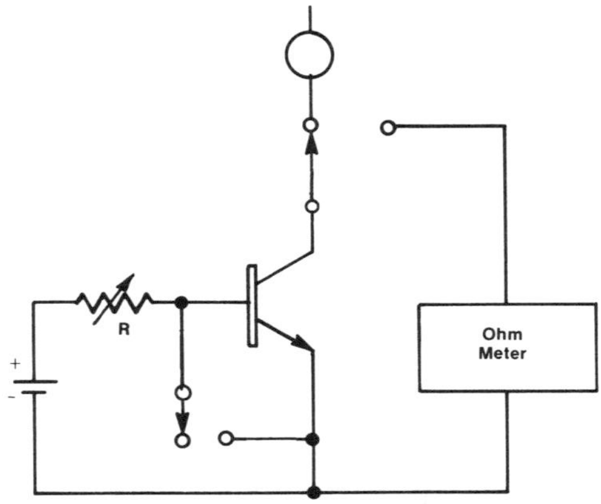

Figure 8-17—Circuit for Measuring Temperature Rise Versus Dissipation for 2N5295

resistance from junction to ambient (Θ_{JA}) was calculated as:

$$\Theta_{JA} = \text{Temperature rise/watts dissipation}$$

In the case of the VN66AF the result was 47.6°C/watt and for the 2N5295 the result was 68°C/watt.

Next, each transistor was checked for temperature rise with a heatsink. The actual heatsink was a small vertical heatsink (see Figure 8-18). With the heatsink, the VN66AF exhibited a temperature rise of 119°C at 10 watts dissipation. This figures out to 11.9°C/watt with the heatsink. This is the junction-to-case plus case-to-ambient (heatsink) thermal resistance, or $\Theta_{JA} + \Theta_{CA}$. This shows that with this heatsink and this particular transistor that 125/11.9 = 10.5 watts can be dissipated at a junction temperature of 125° above ambient. If the ambient is more than 25°C, the dissipation must be reduced accordingly.

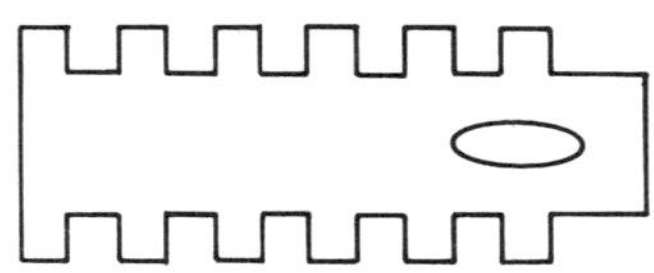

Figure 8-18—Tab Heat Sink

The 2N5295 transistor with heat sink showed a temperature rise of 169°C at a dissipation of 7.5 watts or 169/7.5 = 22.5°C/watt. Since the same heatsink was used for both measurements, one must assume that the junction-to-case thermal resistance (Θ_{JC}) is much higher, on the order of double, in the case of the 2N5295 as compared with VN66AF. Assuming the 2N5295 can be operated up to 200°C as against 150°C for the VN66AF partly offsets the thermal resistance disadvantage of the former.

Plot temperature and current for these two transistors as suggested in Figure 8-19.

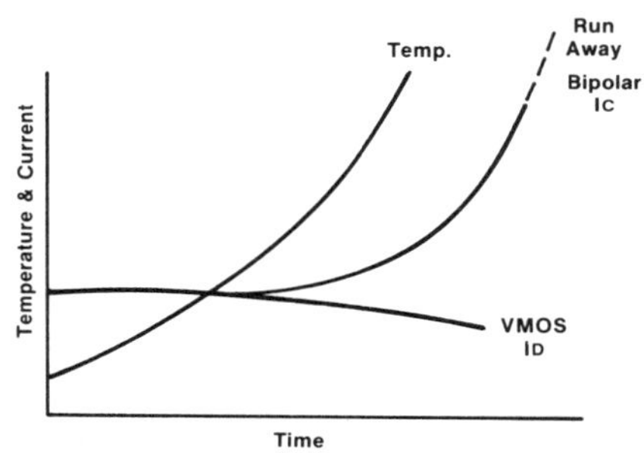

Figure 8-19—Temperature Versus Current for Bipolar and VMOS Transistors

EVALUATING THE EFFECTIVENESS OF AIR FLOW

These thermal dissipation experiments can be expanded in a number of ways. One is to make measurements with several samples of transistors designated by the same part number. Safe operation in any practical system dictates derating to accommodate the worst possible condition.

Another set of experiments can be made with air blowing on the heatsink and of course, different heatsink configurations can be evaluated.

Still other experiments involve runs with mica insulators, with and without the use of silicon grease to improve thermal conductivity, restricting air flow around the heatsink, proximity of heat-dissipating parts such as power transformers and so on.

Testing bright aluminum heatsinks versus black anodized heatsinks leads to interesting evaluations.

STATIC ELECTRICITY

Static electricity was known to the ancient world long before electric current was recognized. It took Benjamin Franklin to prove that the most common form of static electricity, namely, lightning was electrical in nature. Useful electricity is the kind that obeys Ohm's Law, flows around a circuit impelled by an electromotive force, voltage. However, static electricity can be much more spectacular, as in the case of lightning. The interesting thing about static electricity is that it can act over a distance without wires or other conductors. Static electricity can be positive or negative. It can be stored on one's body or in a capacitor. Like charges repell each other, while unlike charges attract each other. Electricity also seeks the ground, where it is neutralized.

Everyone is familiar with the jolt one gets when, after walking on a carpet, one gets too near a grounded object and discharges the accumulated charge with a spark to ground. Static electricity is most evident on a dry day in winter, since dry air is a good insulator and static charges accumulate with little chance to dissipate.

PROJECT: Build a Static Sensitive Light

Two transistors are employed in this project. The input transistor is an enhancement mode VM66 and the output is a 2N2222. When the input transistor conducts, it cuts off the output transistor, turning off the lamp. When the antenna receives a negative static charge, the input transistor cuts off, turning on the output transistor and lighting the lamp. As the antenna discharges, the input transistor drifts into conduction again, turning off the lamp.

The circuit used is shown in Figure 8-20. The finished static-controlled lamp is shown in Figure 8-21 and the internal assembly in Figure 8-22.

Many interesting effects can be created with this static-controlled lamp. Try various insulator rods, say, glass, hard rubber and so forth rubbed on wool or fur; On a dry day in winter, the lamp can be turned on from a distance of several feet. Tie a damp cotton

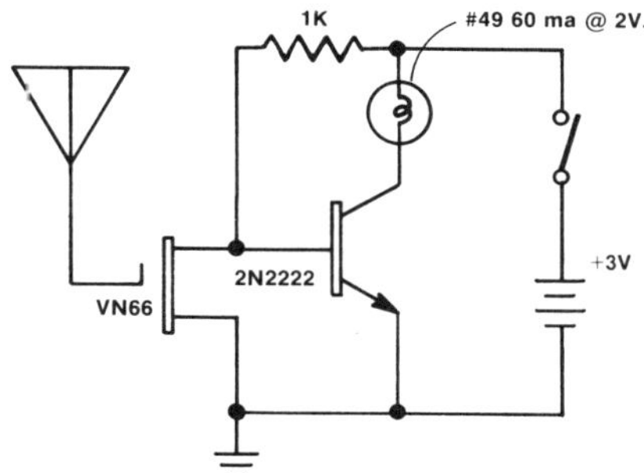

Figure 8-20—Circuit for Static Sensitive Light

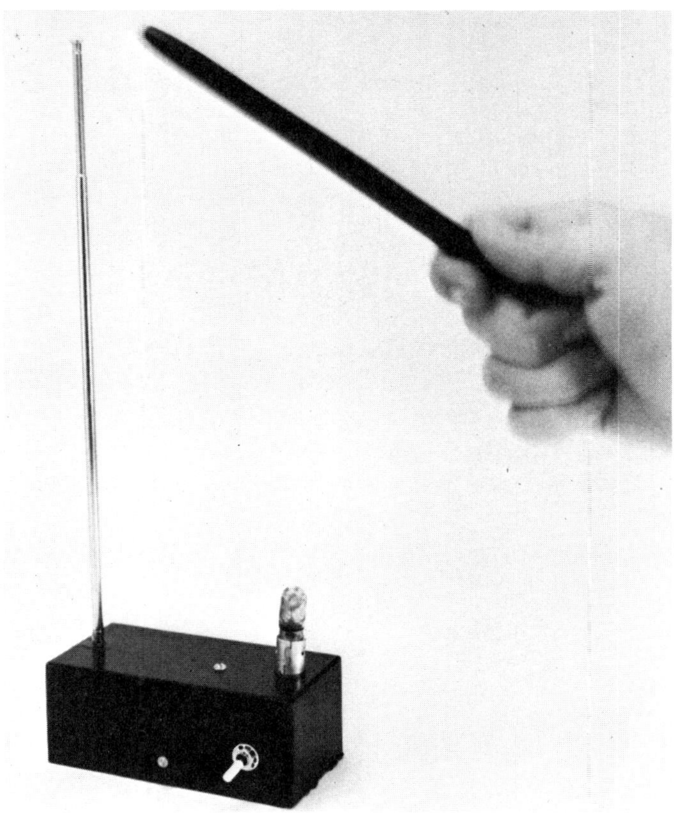

Figure 8-21—Static Sensitive Light Complete

string to the antenna and suspend it with silk threads. It will act as a transmission line for the static charges. When the light is on, blow on the antenna to turn it off. Seek out static charges in a room.

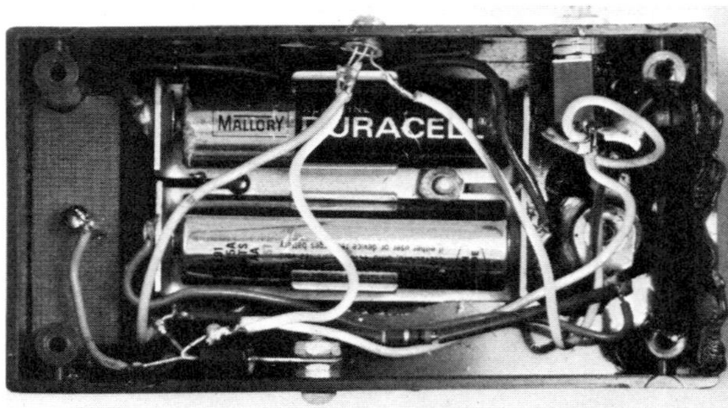

Figure 8-22—Inside Static Sensitive Light

9

PROJECT: Build Your Own Experimental Laboratory

Solid state experiments are generally concerned with circuits. That is, they involve solid state devices and circuit elements such as resistors and capacitors. When solid state devices and circuit elements are combined, circuits are created and experiments can be made by observing performance as circuit elements are changed or varied. Certain assemblies of circuit elements are most useful and time saving in setting up and performing experiments. I will describe below my own version of what I choose to call "Instant Circuits" which enable one to set up an experiment in two or three minutes in many cases.

The "Instant Circuit" system includes a series of component boxes, peripheral devices and a convenient interconnection system. The component boxes house many commonly used resistors, capacitors and inductors. Companion boxes carry transistor and IC sockets, power supplies and meters. All components are accessible by means of pin jacks. The pin jacks are color coded to show values of components, polarities, etc. Interconnections are made by means of color coded wires carrying various combinations of plug and clip terminations. The lead color indicates the length of the particular lead and the type of termination. All these characteristics are aimed at providing a fast effective way of setting up experiments.

CONSTRUCTION DETAILS

Most of the component boxes are the popular 6¼ × 3¾ × 2 inch bakelite boxes with bakelite top panels as shown in Figure 9-1. These boxes are easy to work, will accommodate a wide range of components, are easily arrayed on the test bench and present a professional appearance. The pin-tip jacks used require a ¼ inch hole. Meters require various size holes, drilled with the proper size circular hack saw.

The following list of boxes and their contents have been found the most useful:

Figure 9-1—Undrilled Box and Cover Used for "Instant Circuit" Systems

POTENTIOMETER BOX

Six linear taper carbon (AB) potentiometers (see Figure 9-2) having total resistance values of 50, 500, 5000, 50,000, 500,000, and 5,000,000 ohms. The pin-tip jacks are RMA color coded to identify the potentiometer to which they are connected. For example, the jacks for the 5000 ohm potentiometer are green, black and red (5-0-00). These potentiometers are rated at 2 watts dissipation.

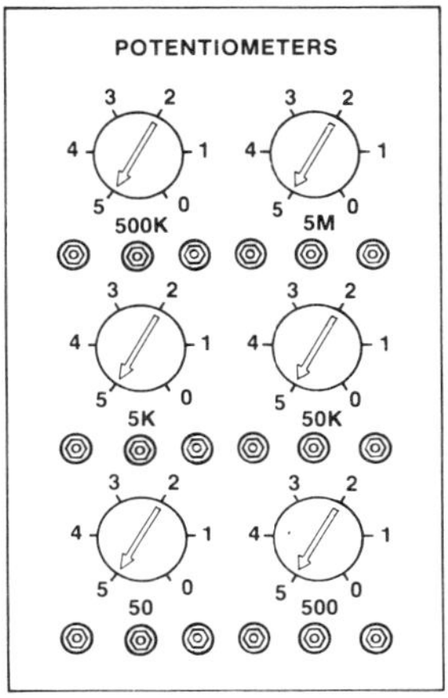

Figure 9-2—Potentiometer Box

VARIABLE RESISTOR BOX

The variable resistor box is similar to the Potentiometer Box but contains 25-watt wirewound potentiometers intended for use where substantial power is to be dissipated as, for example, load resistors for power supplies, power amplifiers and the like. (Ohmite Model H is a suitable potentiometer). Total resistance values of 1; 10; 50; 100; 1000 and 10,000 ohms.

FIXED RESISTOR BOX

The fixed resistor box contains two each of eight different resistance values, namely 100, 470, 1K, 4.7K, 10K, 47K, 100K and 470K. The pairs of resistors have been found most useful, since many circuits require two resistors of the same value. Most circuit uses of these fixed resistors are satisfied with 1-watt 5-percent resistors. Figure 9-3 shows the panel layout and marking. The first three pin-tip jacks in each row are color coded, indicating the value of the resistors in a given row.

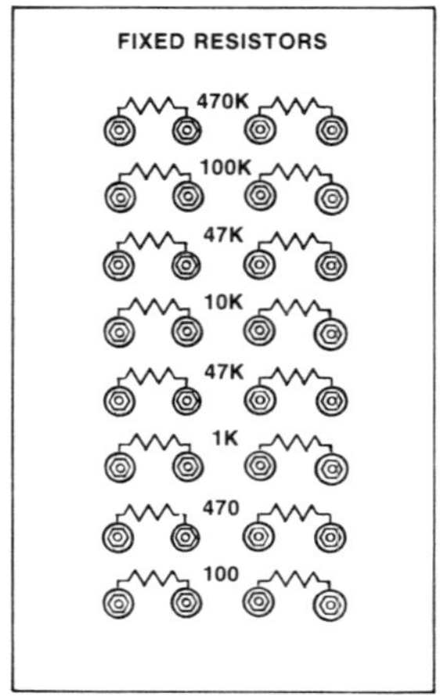

Figure 9-3—Fixed Resistor Box

FIXED CAPACITOR BOX

The same panel drilling is used in the fixed capacitor box panel, but it is marked with capacitor symbols and values as shown in Figure 9-4. Pairs of fixed capacitors having values of .001, .002, .005, .01, .02, 0.5, .1 and .25 microfarads are mounted on the color coded jacks as in the resistor box.

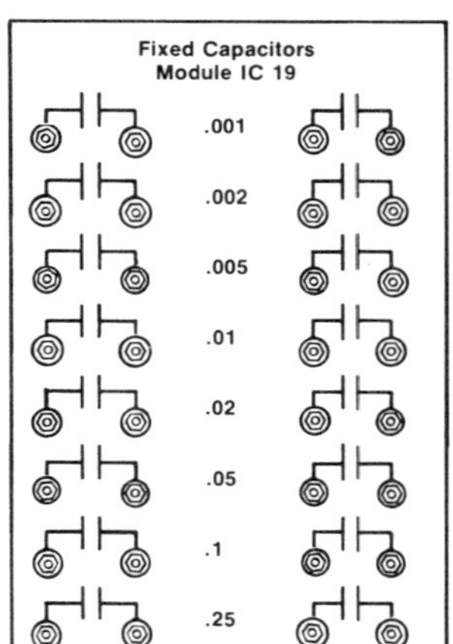

Figure 9-4—Fixed Capacitor Box

BY-PASS CAPACITOR BOX

The By-Pass Capacitor Box is similar to the Fixed Capacitor Box but with pairs of capacitors having values of 0.5, 1.0, 2.0, 5.0, 10.0, 50 and 100 microfarads. Since these must be electrolytic capacitors, the pin-tip jacks are black and red, indicating minus and plus polarities of the capacitors respectively. (Same as Figure 9-4 except for capacitor values).

INDUCTOR BOX

Pairs of inductors having values of 1MH, 10MH, 100MH, 1H, and 10H are contained in the Inductor Box, the front panel of which is shown in Figure 9-5. The same drilling template can be used as was used for the resistor and capacitor box panels, merely omitting 12

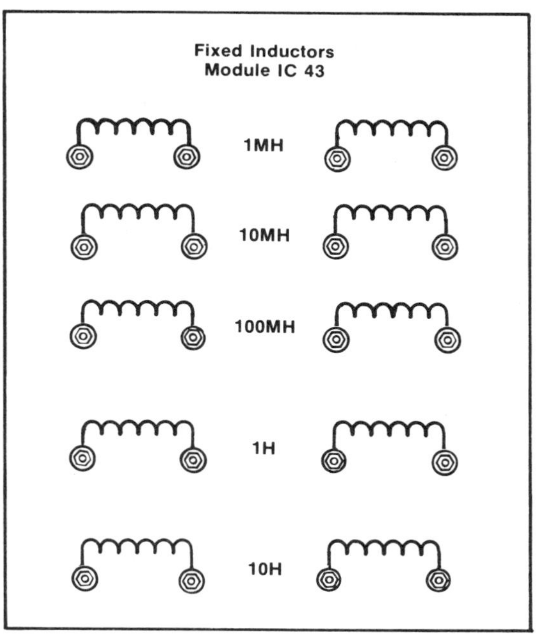

Figure 9-5—Inductor Box

holes. Color coding is used as in the case of the resistor and capacitor boxes.

CURRENT METER BOXES (3)

Current Meter Boxes using 2¼ inch meters are provided with two negative and two positive jacks as shown in Figure 9-6. Three boxes, all containing dc meters, cover the range from 100 microamperes to 10 amperes with the following pairs: 100 microamperes and 1 milliampere, 10 milliamperes and 100 milliamperes, and 1 ampere and 10 amperes full-scale.

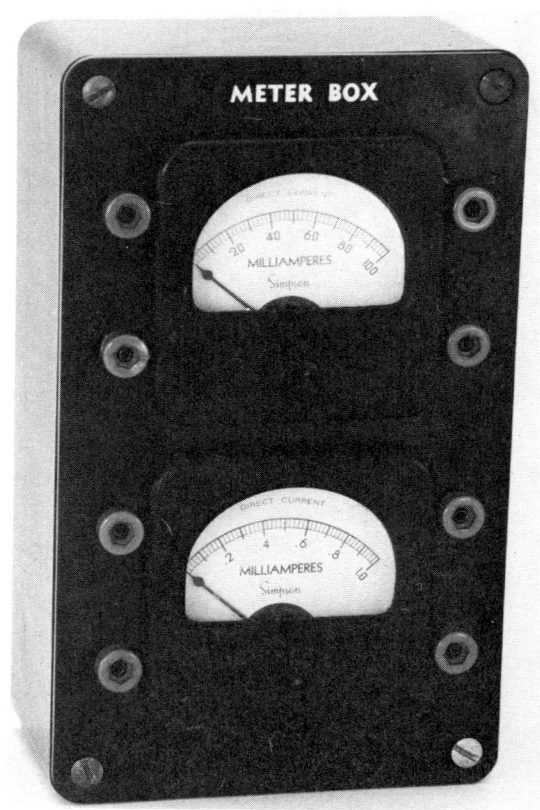

Figure 9-6—Dual Meter Box

Figure 9-6 (cont.)—Wiring of Dual Meter Box

ATTENUATOR BOX

An Attenuator Box with two ten-point attenuators, one made with 100 ohms per step and the other 1000 ohms per step, is shown in Figure 9-7. This attenuator box has many applications, such as a stepped signal source, a stepped feed-back resistor and so forth. Detail of the circuit is shown in Figure 9-8.

TRANSISTOR SOCKET BOX

The transistor socket box carries two transistor sockets and multiple pin-tip jacks on its cover as shown in Figure 9-9. The multiple jacks considerably simplify setting up experimental circuits involving one or two transistors.

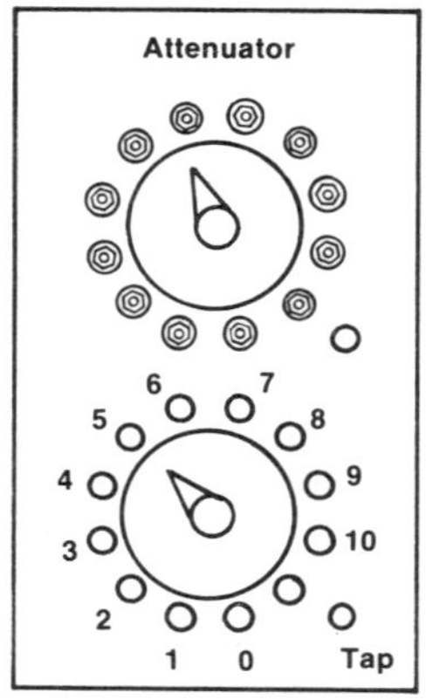

Figure 9-7—Attenuator/Potentiometer Box Panel Layout

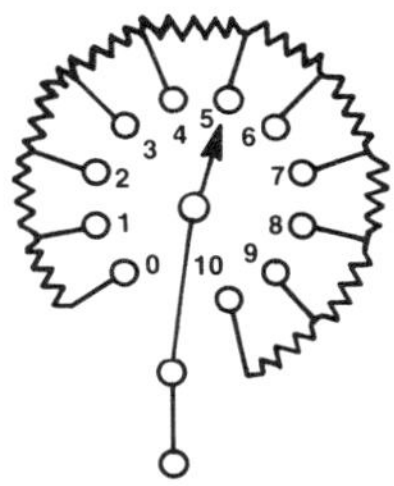

Figure 9-8—Detail of Circuit for Attenuator Box

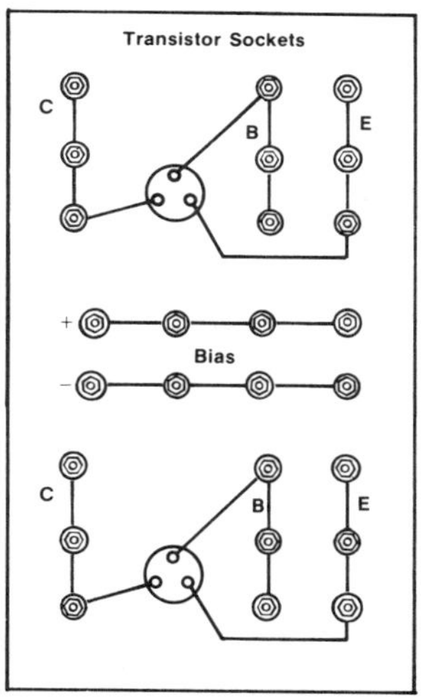

Figure 9-9—Transistor Socket Box

DIP SOCKET BOX

Other socket boxes can obviously be provided as the state of the art changes. For example, the two DIP socket box, as shown in Figure 9-10, is convenient for DIP packages up to 16-pin types. Each socket pin is connected to a correspondingly numbered pin-tip jack.

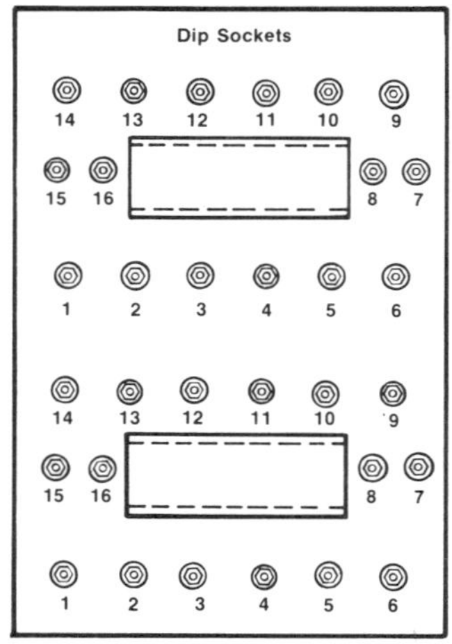

Figure 9-10—DIP Socket Box for Two ICs

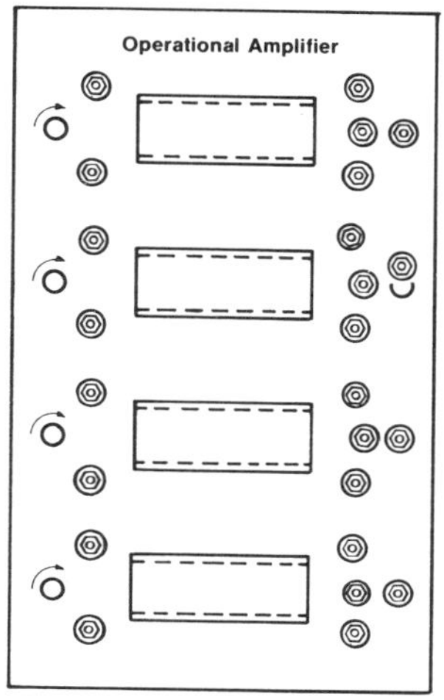

Figure 9-11—Box for Four 741 Operational Amplifiers

OPERATIONAL AMPLIFIER BOX

The 741 operational amplifier is a most useful little amplifier. Figure 9-11 shows a box-top layout with DIP sockets and pin-tip jacks accommodating up to four 741 half DIP operations amplifiers. Figure 9-12 shows the way the jacks are connected, permitting plug-in of input, plus and minus bias voltages and output. A null potentiometer is also provided for each amplifier.

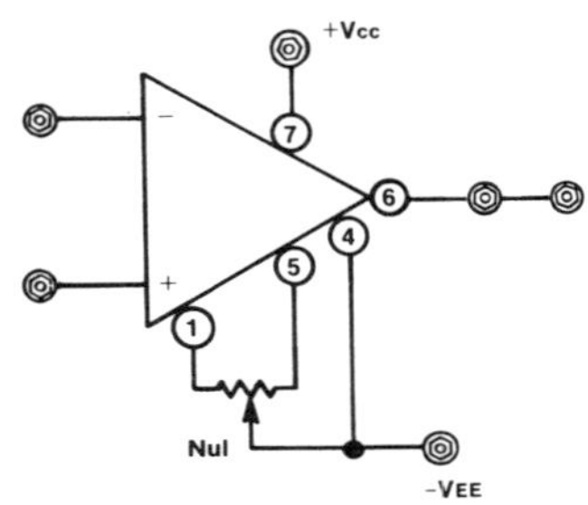

Figure 9-12—Circuit to Show How Jacks Are Connected to Operational Amplifier

SWITCH/LED INDICATOR BOX

The plug-in switch box of Figure 9-13 offers a convenient way to add switches to an experimental circuit. The same box can hold, say, four LED indicators with their built in resistors, useful in indicating circuit states as, for example, IC gates and flip-flops. The internal connections of this box are shown in Figure 9-14. Of course, other switches and indicators can be added, such as push-button switches and incandescent lamps.

VARIATIONS POSSIBLE IN CIRCUIT BOXES

The boxes decribed and shown above have been chosen since they have been the most used in my own experiments. Incidentally, more than one box of a given type may also be useful and many modifications are possible. Some types of experimenting may indicate potential usefulness of many other types of circuit boxes. A little experience with this concept will show that a great deal of time can be saved by their use. One feature of the system is that

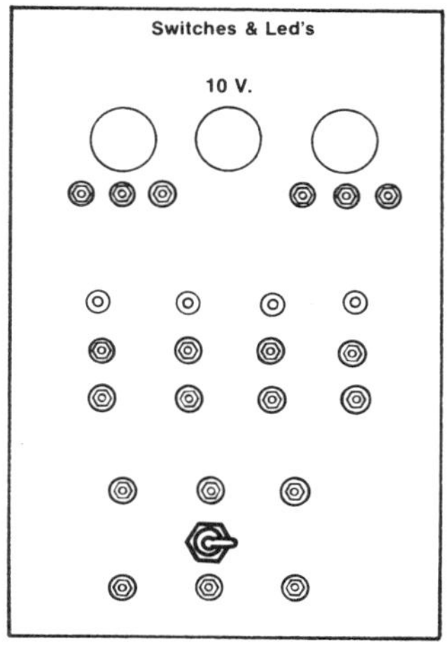

Figure 9-13—Switch/LED Indicator Box

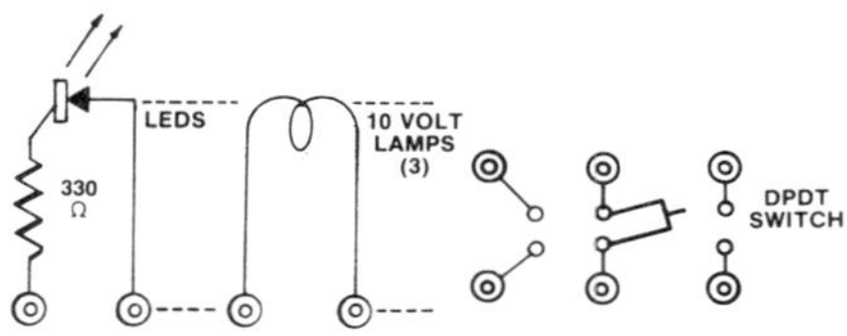

Figure 9-14—Internal Connections for Switch/LED Box

experiments can be performed or demonstrations made at another place by selecting a few boxes, carrying them to where you want to make a demonstration and quickly setting them up.

THE INSTANT CIRCUIT LEAD CONCEPT

To obtain the full benefit of the Instant Circuit concept a lead system was devised which has proved most compatible. The leads are of various lengths and the length is indicated by the color of the wire and, in some cases, by the color of the pin-tip plug body. Leads of a given length are provided with the same terminations, for example, the yellow lead is 5 inches long and has a plug on one end and a small alligator clip on the other end. All one has to do is to determine the length of lead and the terminations required for a given connection and then pick up the proper lead by choosing the color meeting the requirement. The system is quickly learned and, as an aid, the lead lengths progress in accordance with the RMA color code. (See Figure 9-15 for sample).

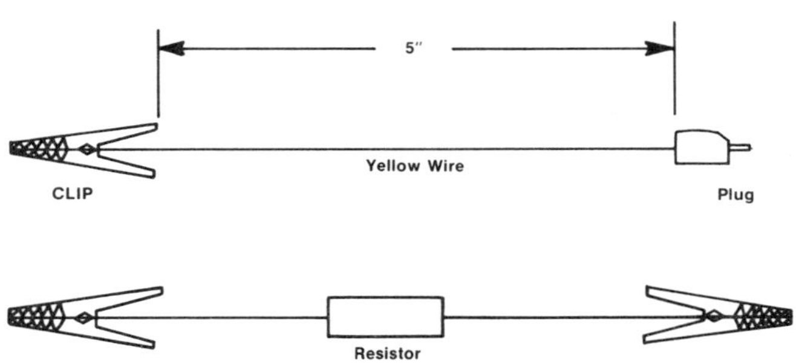

Figure 9-15—Typical Leads and Components

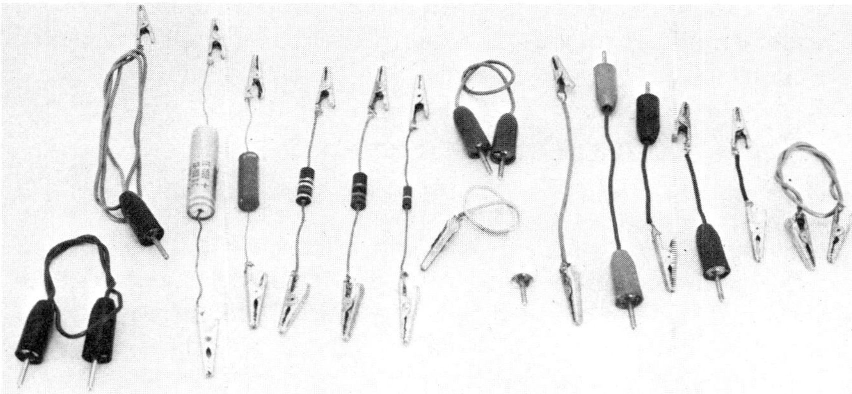

Tabulation of Lead Specifications

1. Clips on both ends

Length	Color
1 inch	black
4 inches	orange
8 inches	blue
16 inches	gray

2. Clip on one end, plug on other end

Length	Color
2 inches	brown
5 inches	yellow
10 inches	violet

3. Plugs on both ends

Length	Color
3 inches	red
6 inches	green
12 inches	white

A good working complement of leads would be, say, ten of the shorter lengths up to and including 6 inches green and about six of the longer ones.

CLIP-ON DISCRETE COMPONENTS

While the component boxes interconnected by means of the color coded leads provide most of the circuit components required in the average experiment, a supply of discrete components terminated by small alligator clips adds flexibility to the system. Two each of the most used resistor values, such as 100 ohms, 1K, 10K, 100K and 1 meg., and one each of a few capacitors such as .01 mfd, 0.1 mfd and 1 mfd are suggested. An example is included in drawing Figure 9-15.

USING INSTANT CIRCUITS

Figure 9-16 is a drawing of a typical experimental set-up using the Instant Circuit System. Figure 9-17 is the actual schematic of the set-up of Figure 9-16. On a test run it took 12 minutes to connect up this circuit and make it operational including drawing the schematic diagram and selecting the boxes to be used.

NOTE ON TEST EQUIPMENT

Experimenting in solid state electronics, as in other fields, is pretty well defined by the test equipment used. There was a time when an experimenter might very well build most, if not all, of his test equipment himself. However, over the years, much standard equipment has been developed to such a high degree that it is quite impractical to attempt to build your own. On the other hand many experiments may require special equipment which is not readily available or is only available in unnecessarily sophisticated form. An experimenter can, for example, build his own power supplies, amplifiers, active filters, transistor checker, and so on.

PROJECT: Build a Microvolt/Milliammeter

The operational amplifier is widely used in test equipment. One simple but very useful instrument is a combination microvolt/milliammeter. With two 741 operational amplifiers, a 1 milliampere

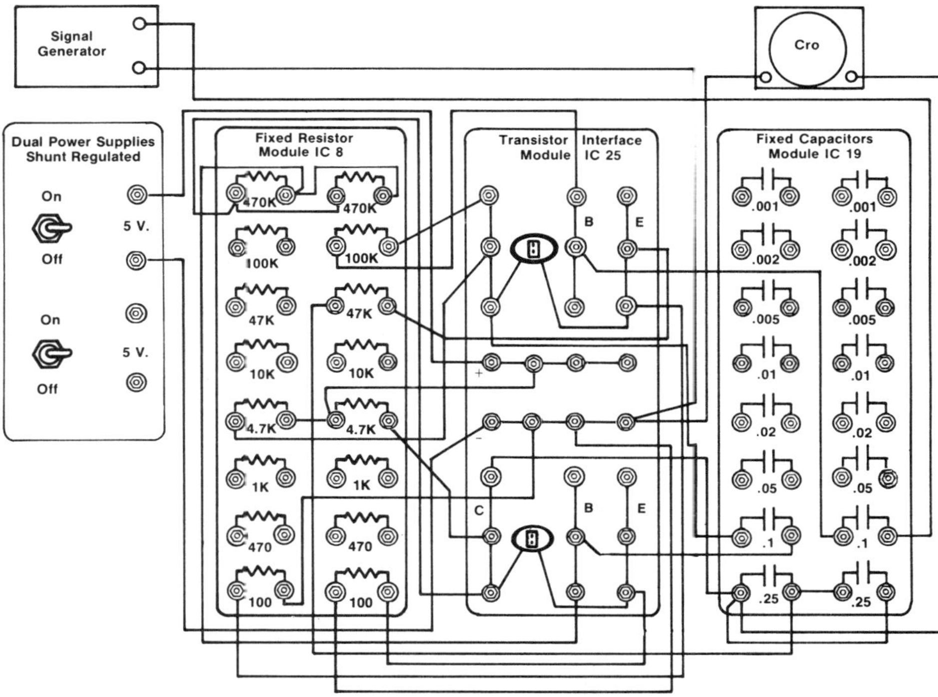

Figure 9-16—Set-Up Using Instant Circuit System

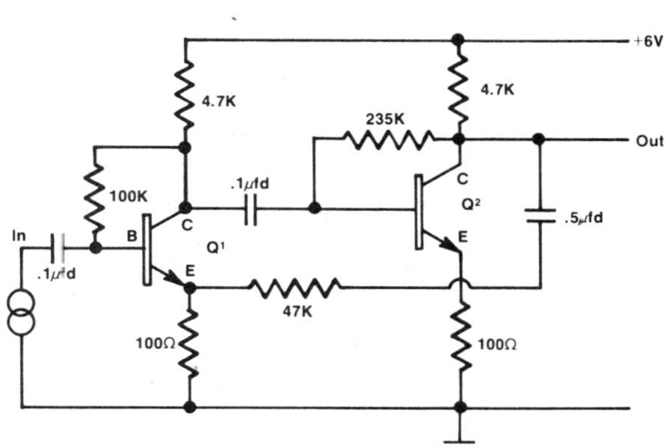

Figure 9-17—Schematic of Set-Up Using Instant Circuit System

meter, a few resistors, a potentiometer and a double-pole double-throw switch, this versatile meter can be constructed in a few hours. With the circuit and constants as shown in Figure 9-18, full-scale

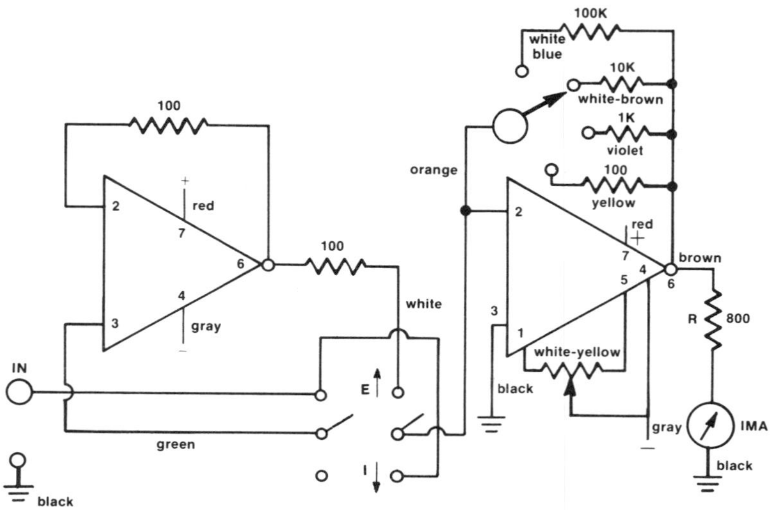

Figure 9-18—Schematic of Microvolt/Milliammeter

Figure 9-19—Microvolt/Milliammeter Completed

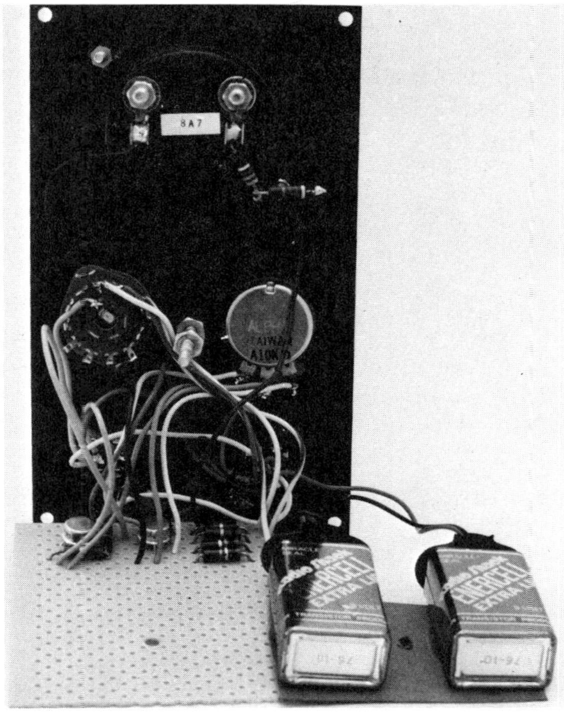

Figure 9-20—Microvolt/Milliammeter Inside

current ranges of 10 and 100 microamperes and 1 and 10 milliamperes and voltage ranges of 1, 10, 100 millivolts and 1 volt can be provided. The voltmeter input impedance is of the order of 1 to 2 megohms per volt and the current-meter impedance is substantially zero. The ability to measure microamperes with substantially zero voltage drop is a unique capability not possible without such assistance as the operational amplifier. The operational amplifier provides a virtual ground across its inputs with substantially no voltage drop or a null.

Construction Hints: The amplifier and resistors can be conveniently mounted and wired on a small perf board with 0.1-inch hole spacing. For all but most intensive use, a pair of 9-volt batteries provide a convenient bias source. The calibrating resistor R is chosen so that the 1 ma meter reads full scale when the amplifier output is 1 volt. I would recommend, as with all battery operated equipment, that the flashing ON indicator be included. The standard 6¼ × 3¾ × 2 bakelite box is just about ideal for this instrument. Figure 9-18 is the complete schematic circuit diagram, including suggested lead color coding to facilitate final assembly. Figures 9-19 and 9-20 are photographs of the finished meter and its internal construction.

10

Projects and Experiments with Linear Integrated Circuits

Solid state technology has brought out a large number of highly developed amplifying devices. The most versatile is the operational amplifier. In order to more fully appreciate this most useful device a number of significant experiments are outlined below.

EXPERIMENTING WITH OPERATIONAL AMPLIFIERS

An operational amplifier may be defined as an amplifier having a high gain (open loop gain) and stable characteristics so that its actual gain (closed loop gain) is essentially determined by a degenerative feedback network. The symbol and basic circuit for an operational amplifier are shown in Figure 10-1. The closed loop gain of such a circuit is R_2/R_1. The frequency characteristic of a stable operational amplifier is shown in Figure 10-2. It will be noted that maximum gain is at zero frequency, in other words, with direct current. That the gain remains constant at very low frequencies, after which it falls uniformly at a rate of 6DB per octave, is the prerequisite gain versus frequency characteristic of a stable operational amplifier. The gain-determining impedances set an operating charcteristic as shown by the horizontal dotted line. The frequency response will be uniform to the point of intersection with the gain versus frequency curve, from which point it will decrease 6 DB per octave.

Since operational amplifiers are applied in many different ways, such amplifiers have been developed with a variety of characteristics. Thus, selection and/or testing should be done in view of the intended use.

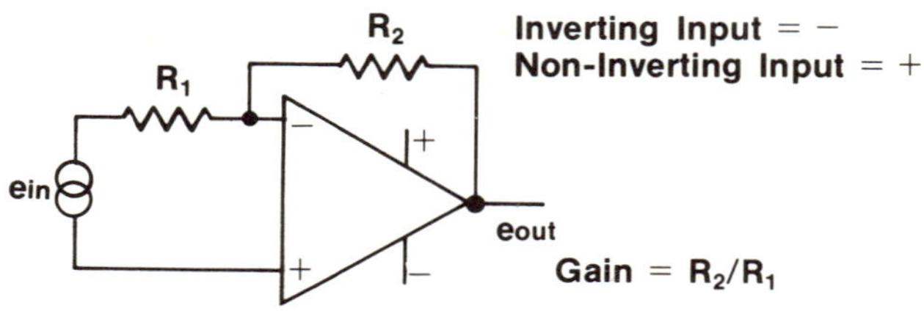

Figure 10-1—Symbol and Basic Circuit for Operational Amplifier

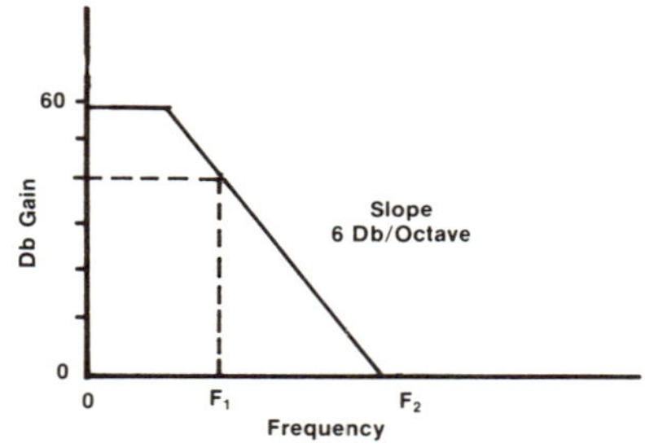

Figure 10-2—Frequency Characteristic of a Stable Operational Amplifier

MEASURING OP AMP CHARACTERISTICS

The dc gain of an operational amplifier is generally difficult to measure, since on open loop the amplifier snaps from bottoming in one direction to bottoming in the other. Hysteresis makes it almost impossible to hold the output stable at an intermediate point. One way to approximate the dc gain measurement is to use a feedback capacitor C so that gain at audio frequencies is reduced to a low value. Figure 10-3 shows a simple set-up for making such an experiment. The null adjustment must be made as accurately as possible. Measuring E_{OUT} and calculating E_{IN} provides the basis for the gain measurement. Generally, the figure obtained will be close to the actual dc gain.

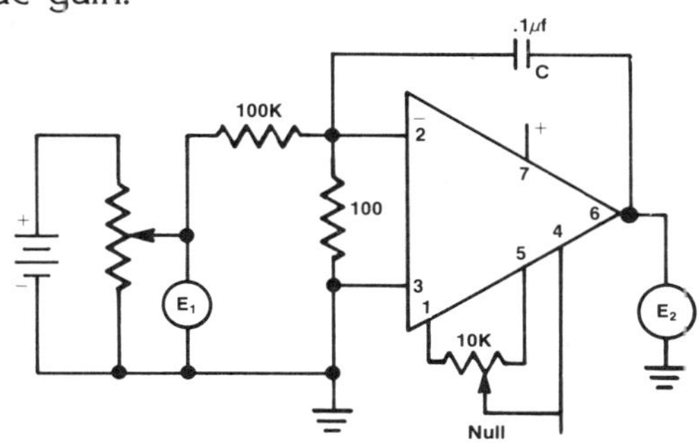

Figure 10-3—Set-Up for Measuring dc Gain

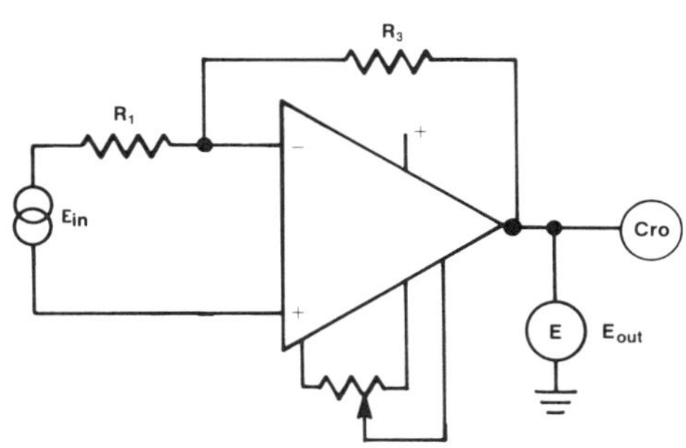

Figure 10-4—One Set-Up for Measuring Gain Versus Frequency

The circuits in Figures 10-4 and 10-5 show set-ups for measuring gain versus frequency. The wider the frequency response desired, the lower the gain must be. In making such measurements on a 741 op amp, the following data was obtained, in which the frequencies at which the gain was down 3 DB were recorded:

Gain	− 3 DB Frequency
1000	1500 Hz
500	3000 Hz
200	20 K Hz
100	40 K Hz

If more gain is required for a given frequency response, two or more op amps can be cascaded. Stability at dc will often dictate using ac (capacitor) coupling between succeeding amplifiers.

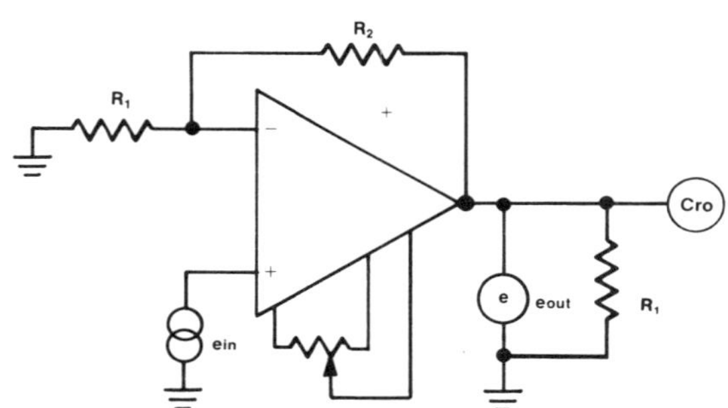

Figure 10-5—Another Set-Up for Measuring Gain Versus Frequency

EXPERIMENTING WITH R, L, C FEEDBACK

Internally compensated op amps are stable under a wide range of feedback conditions. This makes them useful in applications where shaped frequency response is required as in the case of active filters. Figure 10-6 shows two simple forms of selective feedback. Form A is a notch filter causing a deep dip in gain, while form B does the opposite, causing a peaked gain characteristic. An almost endless variety of response curves can thus be generated. They range from simple rising or falling response with frequency to band-pass, band-elimination and so on. The fact that amplification takes place along with frequency discrimination gives this type of circuit a substantial advantage over the older passive component configurations.

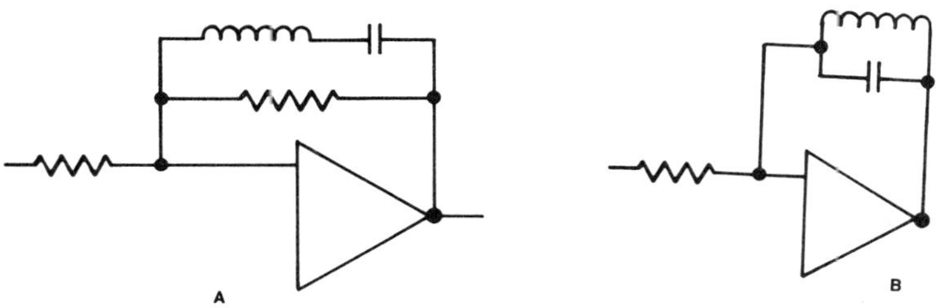

Figure 10-6—Two Simple Forms of Selective Feedback

DEALING WITH OFF-SETS IN OP AMPS.

Since op amps respond down to zero frequency (dc), they are subject to two effects which are potential sources of trouble; the offsets. In an ideal op amp, the input is perfectly balanced both for voltage and current. In other words, with no applied input voltage or current, the output sits at zero volts. This is never the case in an actual amplifier. The input exhibits an offset voltage and an offset current. These offsets are amplified by an amount equal to the closed loop gain of the amplifier. The voltage offset appearing as an amplified voltage at the output and the current as a voltage drop in the feedback resistor. In order to overcome this problem, terminals are provided for connecting a nulling potentiometer, the proper adjustment of which will bring the offset effect to zero. Unfortunately, this is not a perfect solution to the problem, since the offset is generally temperature dependent and one nulling adjustment will not hold over a wide range of temperature. Chopper stabilization is often employed where offset is a critical problem.

PROJECT: Build High Fidelity Amplifiers

The most useful type of power amplifer employs two power transistors, one passing the positive signal excursions and the other the negative. There are a number of ways in which this can be done. In the days of the vacuum tube, it was done with transformers since tubes are single polarity (NPN) devices. However, with transistors of both positive (NPN) and negative (PNP) polarities the transformers can be done away with and directly connected circuits can be utilized. While there are a number of directly connected circuits possible, the easiest to experiment with is the complementary symmetry circuit shown in elemental form in Figure 10-7.

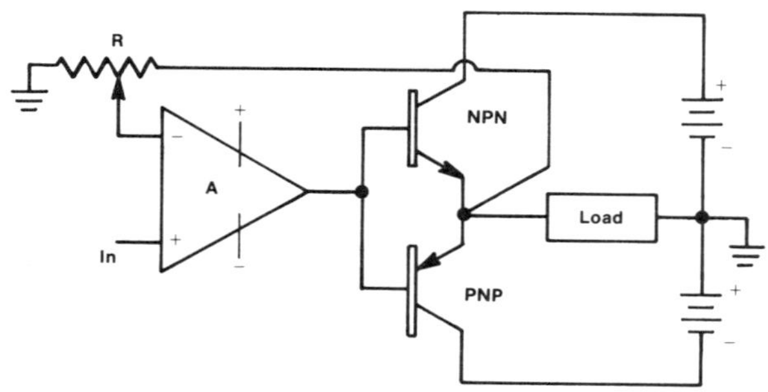

Figure 10-7—Simple Circuit for a High Fidelity Amplifier

The complementary symmetry circuit has two basic requirements; first, a pair of matched transistors, one NPN and the other PNP; and second, a balanced power supply; equal positive and negative voltages with respect to common or ground. The simple basic circuit of Figure 10-7 includes a driver amplifier, A, and a gain control, R, acting as a feedback and input resistor for op amp A. An important feature of this circuit is that the feedback is taken from across the load. This is especially advantageous when this load is a loudspeaker.

Experiments with the circuit of Figure 10-7 will quickly demonstrate two basic problems. First, the typical op amp has a limited current output and will be unable to drive the power transistors to a level of more than a few watts of output. Second, since transistors require a fraction of a volt input before they start to conduct, there will be intervals between the conduction in the positive direction and conduction in the negative direction as illustrated in Figure 10-8. This is called "cross-over distortion" and is

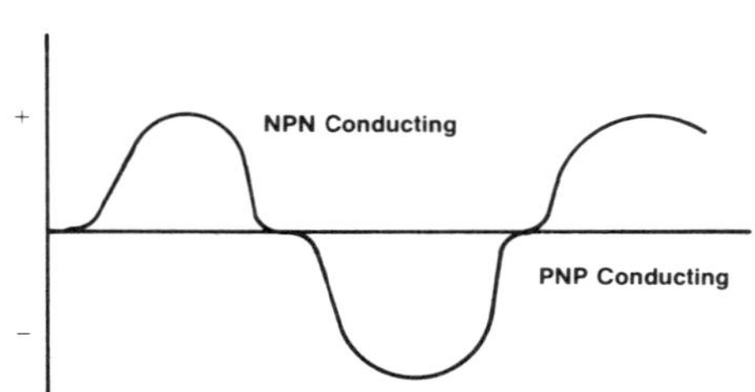

Figure 10-8—Intervals Between Positive and Negative Conduction

especially noticeable at low levels of output sound. Eliminating this cross-over distortion is not difficult, but unless care is taken, such elimination can readily and substantially decrease the efficiency of the amplifier. Decreased efficiency can, in turn, lead to excess heating of the transistors.

EXPERIMENTS WITH FEEDBACK

Figure 10-9 through 10-13 are five permutations of complementary symmetry transistors power amplifier circuits. An experimental set-up in each case should include a source of variable frequency audio signals, an output voltmeter, a cathode ray tube for observing output waveforms, and if available, a distortion meter, preferably one that reads intermodulation distortion directly.

The experiment to be performed, based on Figure 10-9, is to observe two effects. First, the difference in the feedback effect between connection A, from the op amp output, and connection B, from the load. The feedback from the load, particularly when the load is a loudspeaker is most effective, since it not only counteracts distortion in the power transistors but also acts to dampen the loudspeaker transients.

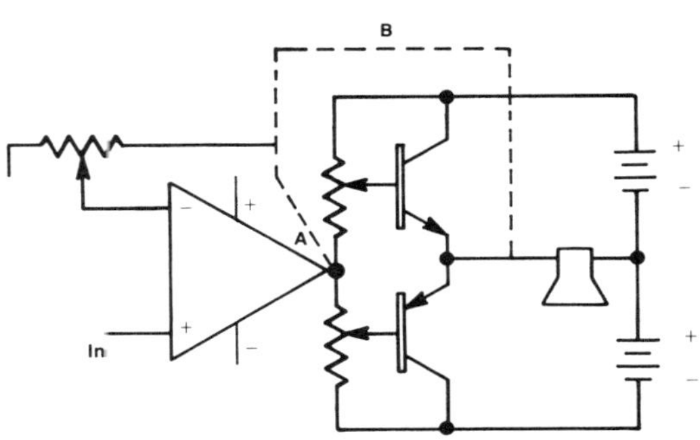

Figure 10-9—Basic Transistor Power Amplifier Circuit

The second experiment is to find out what adjustments of the two base-connected potentiometers will eliminate cross-over distortion. To start, the variable contacts should be set at the ends connected to the amplifier output. They should be advanced a little at a time, observing the cross-over distortion until it just disappears or a sudden drop occurs in the IM distortion reading.

Adding Darlington Drivers

Figure 10-10 shows how Darlington drivers can be added. These will provide greater power output from the power transistor pair for a given op amp. The power transistors require substantial driving current to provide more than a fraction of a watt output, which cannot be provided by the usual op amp. The Darlingtons will supply substantial current gain, thereby increasing the power output. In this circuit, the feedback from the load is very important, as otherwise, the Darlington combination can cause instability in the circuit, evidenced by a thermally induced drift. This can be checked out by returning the feedback to the amplifier output.

It will be seen that the feedback has the same phase at the load as it has at the amplified output, since the power transistors are connected as emitter followers.

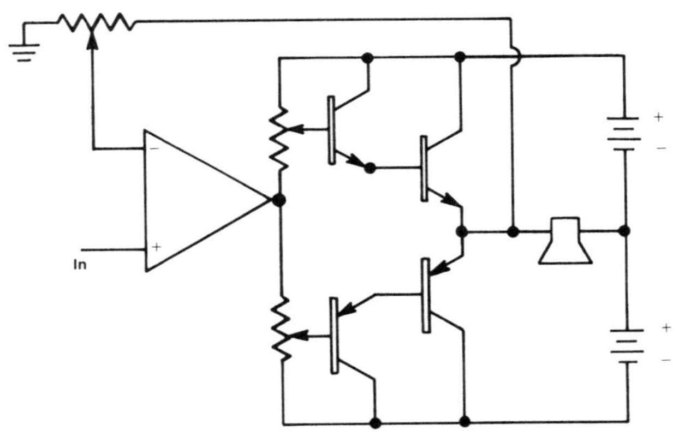

Figure 10-10—Circuit with Added Darlington Drivers

Adding a Single Transistor Driver

Figure 10-11 shows how driving current can be supplied by means of a single transistor connected between the op amp and the power transistors. The collector and emitter resistors are shown as variable. Experimenting with them will determine what the emitter resistor must be to enable the op amp to exert control. The base and emitter voltages differ by only a fraction of a volt so that if the driver transistor emitter is connected directly to the negative voltage source, the op amp will probably be unable to turn it off. However, if this resistor is too large, the drive transistor will be unable to pass sufficient drive current to the PNP power transistor. Similarly, if the collector resistor is too large, it will not pass sufficient drive current to the NPN power transistor. On the other hand, making these resistors too small (in resistance value) will force the drive transistor to dissipate more power than is necessary for efficient operation of the circuit and will make it necessary to use a higher power transistor for the drive transistor.

It should be noted that the feedback from the load in this circuit goes to the plus terminal of the op amp since the driver transistor is connected to invert the phase. One might connect the drive point to the emitter of the drive transistor, in which case there would be no change of phase and feedback would be to the negative input (inverting) of the op amp.

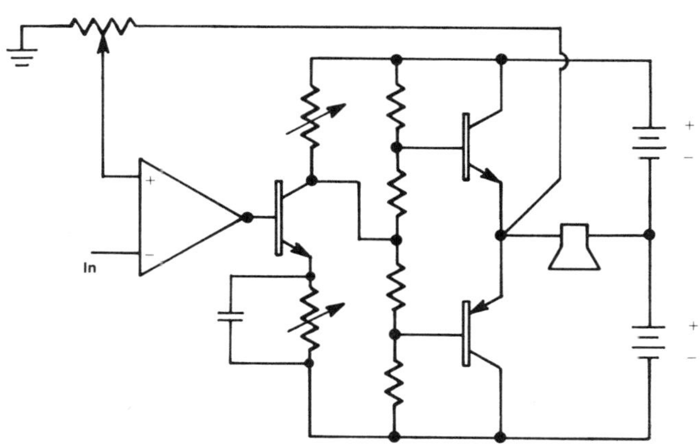

Figure 10-11—Circuit with Added Single Transistor Driver

Power Amplifier Refinements

In addition to the basic concepts outlined above there are many refinements which an experimenter can try in the power amplifer art. One refinement is suggested in the circuit of Figure 10-

12. Here, two diodes are used to provide the crossover eliminating offset. One problem with the simple reistor circuits (See Figure 10-9, 10-10 and 10-11) is that as the temperature of the power transistors rise, the optimum off-set decreases. Thus, if the diodes are thermally coupled to the power transistors, as in Figure 10-12, they will provide the better (decreasing) offset with increased temperatures.

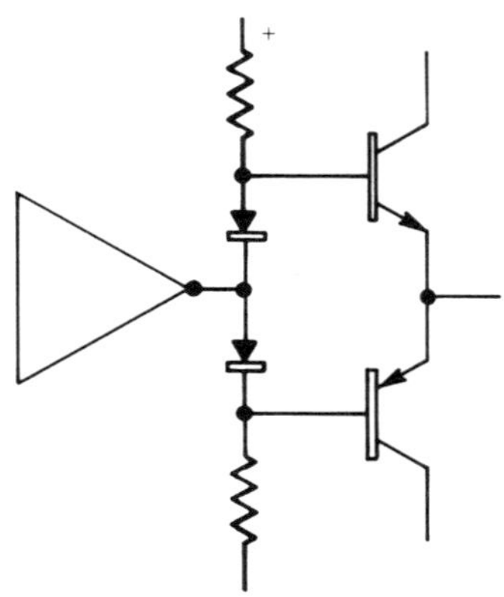

Figure 10-12—Circuit with Thermally Coupled Diodes

Another refinement, one which becomes increasingly important as the output power capability is increased, is overload or short-circuit protection. Resistors in series with the power transistor emitters, as shown in Figure 10-13 are one way to provide overload protection. In addition, a sharper cutout can be provided by comparing the voltage across these resistors with a reference voltage and reducing or cutting off base drive with a feedback amplifier whenever the voltage across these resistors exceeds the reference voltage. The voltage across the emitter resistor is, of course, a measure of the transistor emitter current, which must be limited for safe operation. The load limits the current in normal operation, but a short across the load removes this limiting effect and effectively places the power supply voltage directly across the transistor.

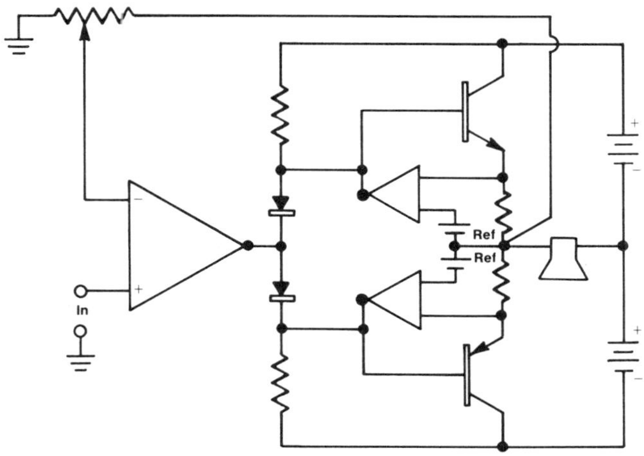

Figure 10-13—Circuit with Short Circuit Protection

BALANCING POWER TRANSISTORS

Ideally, both transistors of a complementary symmetry pair are perfectly balanced. That is, at any input current to a base, the resulting emitter currents are the same. Unfortunately it is not easy to make perfectly matched NPN and PNP power transistors. Feedback from the load tends to correct imbalance, but finally, the output will be limited by the transistor of the pair having the lower current gain (B).

How to Obtain Matched Power Transistors

One way to at least partially overcome the problem of matching power transistors is to use two power transistors of the same type in the output; either two NPN or two PNP transistors. Then it is relatively easy to select two matching transistors of the same type.

The basic circuit for doing this is the quasi-complementary symmetry circuit as shown in Figure 10-14. Here, the output transistors are both NPN and the driver transistors are a complementary pair. It is relatively easy to find a matching pair of driver transistors.

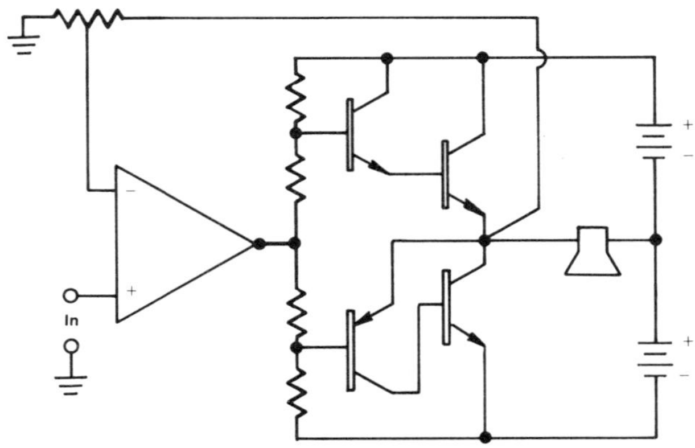

Figure 10-14—Circuit with Matched Power Transistors

POWER SUPPLIES FOR POWER AMPLIFIERS

As the output power of any power amplifier in increased, the demand on its power supply is also increased. For example, a 20-volt power supply for a one-watt amplifier (assuming 50 percent efficiency in the amplifier) is required to supply only 100 milliamperes of current. However, at 20 watts output, the power supply must supply 2 amperes of current which is quite another story. It is merely pointed out here that as the power to be derived from a power amplifier is increased, the requirements of the power supply are likewise increased. This means extra power supply capacity, increased rectifier ratings and greatly increased filter capacitor capacity and/or regulation.

Choosing a Single or Dual Power Supply

Figures 10-7 through 10-14 shown power output circuits using dual power supplies. These circuits have some advantages, but it is also possible to power these same circuits, with one significant change, from a single power supply, as shown in Figure 10-15. This is made possible by the use of capacitor C. However, it must be a large capacitor. If the system's impedance is to equal 8 ohms at 50 cycles C must be 400 microfarads. Another requirement is that the driver amplifier must be able to supply an offset equal to half the power supply's total voltage. There are other complications with the single power supply type of circuit which the experimenter will discover and will have to deal with if he insists on using a single power supply.

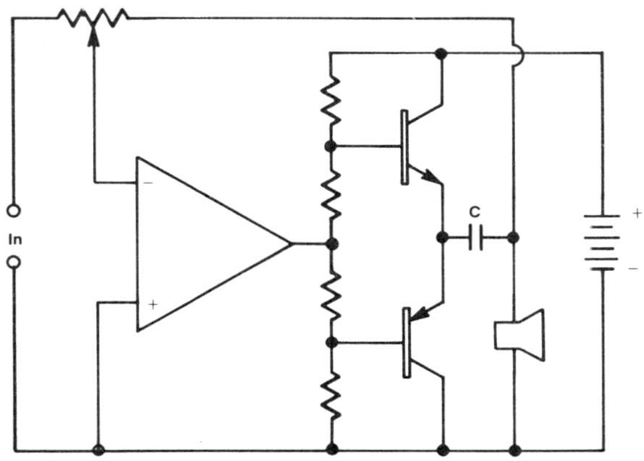

Figure 10-15—Powering the Circuits from a Single Power Supply

PROJECT: Harmonic Analyzer

Harmonic analysis is a "must" in working with high fidelity equipment. However, harmonic analyzers which measure each individual harmonic are complex and expensive. On the other hand, the measurement of total harmonic distortion (THD) is much easier. The result is that, many times, equipment is measured and rated in THD terms rather than in individual harmonics. This does not alter the fact that individual harmonics have significance not disclosed by THD measurements. The second and third harmonics are generally the most significant. For example, a push-pull amplifier may be so well balanced as to proivide a very low second harmonic, but it may be overloading on positive and negative excursions so as to produce a relatively large third harmonic.

A simple and inexpensive selective amplifier tuned to a single frequency can be used as an effective instrument for measuring individual harmonics. It works this way. A very selective amplifier tuned to 5000 cycles, for example, can be used with a source of low distortion variable frequency. The amplifier has an output meter in order to make measurements. First, the device under test is fed 5000 cycles and the amplifier gain is adjusted to read full scale on its output meter. Then the input signal frequency is changed to 2500 cycles so that the amplifier is selecting the second harmonic of the input. The amplifier gain is increased by known increments until a meaningful reading is obtained on the output meter. The amplitude of the harmonic is then the percentage read on the meter divided by the increased gain increments. For example, suppose the meter reads −6DB and the gain has been increased 20 DB, the second harmonic is −26 DB below the fundamental. The third harmonic is measured by changing the input frequency to 1667 cycles and repeating the procedure and so on.

This method of measuring harmonics assumes flat gain characteristics in the equipment being examined, not unreasonable in high fidelity technology.

The project to be described employs a 741 operational amplifier with a high Q inductor tuned to 5000 cycles by means of a capacitor connected as the feedback means. This results in a highly selective amplifier peaking at 5000 cycles. The amplifier output is indicated on a VU meter reading DB above and below 0. In the calibration step, the meter is to read 0 so that the scale reads directly in DB below 0. Two input attenuators are provided. One is a continuously variable potentiometer for setting over all gain. The second is a step attenuator providing attenuation of 10 DB per step. The 0 step is at maximum attenuation and is used in calibrating the 5000 cycle output. When measuring a harmonic, the attenuator output is increased until the amplifier output meter reads between 0 and −10 DB. The harmonic is then below the fundamental by the meter reading plus the attenuator setting, for example, −6 DB −30 DB or a total of −36 DB.

CONSTRUCTION DETAILS OF HARMONIC ANALYZER

The analyzer was built in a 6¾ × 5½ × 2-7/16 inch Bakelite box. An internal power supply was provided.

The complete circuit diagram is shown in Figure 10-16: R_1 provides the input calibrating gain control; R_2 through R_6 provide the step attenuator; the amplifier is a 741 op amp; the inductor is a 1-henry toroid; the tuning capacitor is 840 micromicrofarads composed of 330 and 510 mmf in parallel, plus a fine adjusting 50 mmf variable; the meter is a Micronta VU meter. The meter is protected by two back-to-back LED's. Figure 10-17 is the circuit of the power supply and Figure 10-18 is a drawing of its assembly. Figure 10-19 is a photograph of the finished analyzer while Figure 10-20 is a photograph of the internal assembly and wiring. Figure 10-21 shows drilling dimensions for the front panel.

The parts are rather obvious and are generally available at Radio Shack, Lafayette and other parts suppliers. The meter used in the model built to check out the design is a Radio Shack Micronta VU meter. The case has a cover which is a full 6¾ × 5¼ inches. If a Calectro phenolic chassis box is chosen, the cover is only 6½ × 5 inches and the drilling pattern must be modified by subtracting ⅛ inch from all dimensions referenced to outside edges of the panel. The resistors required by the attenuator are best provided by combining series and parallel ½ and ¼-watt resistors in various combinations until acceptable values are obtained. Some will come off easier than others. Final values within, say, 5 percent should be close enough in any case. Use your own judgement.

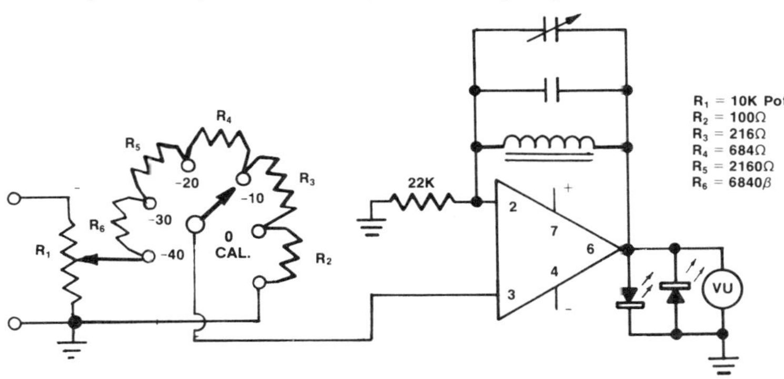

Figure 10-16—Circuit Diagram for Harmonic Analyzer

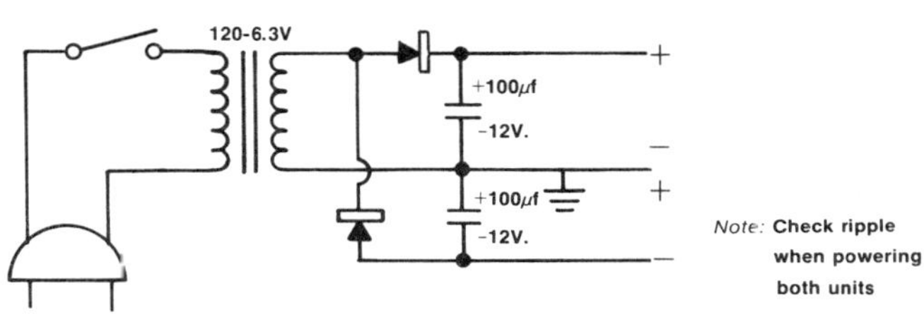

Figure 10-17—Circuit of Power Supply for Harmonic Analyzer

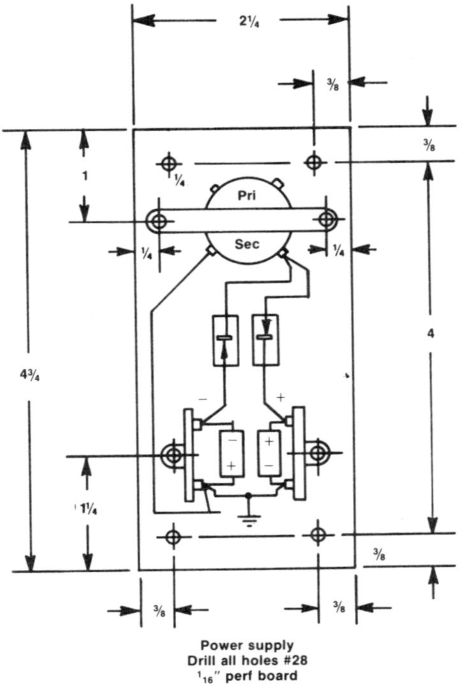

Figure 10-18—Assembly of Harmonic Analyzer Power Supply

Figure 10-19—Harmonic Analyzer Completed

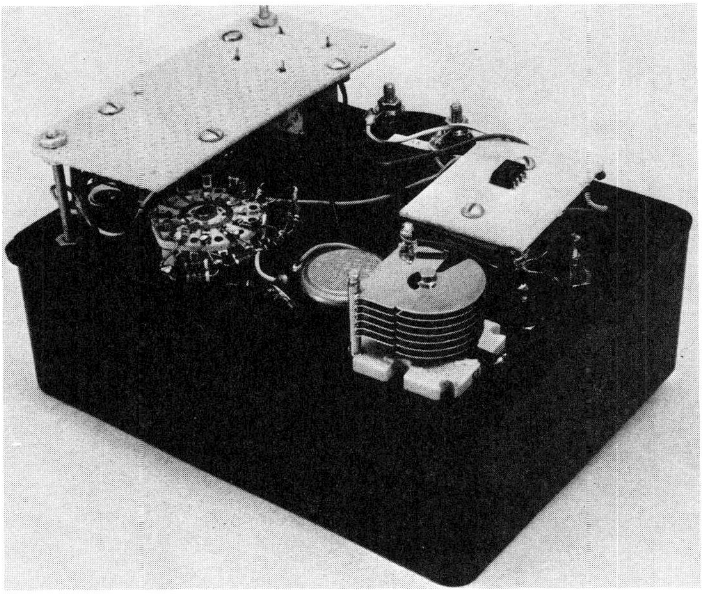

Figure 10-20—Inside of Harmonic Analyzer

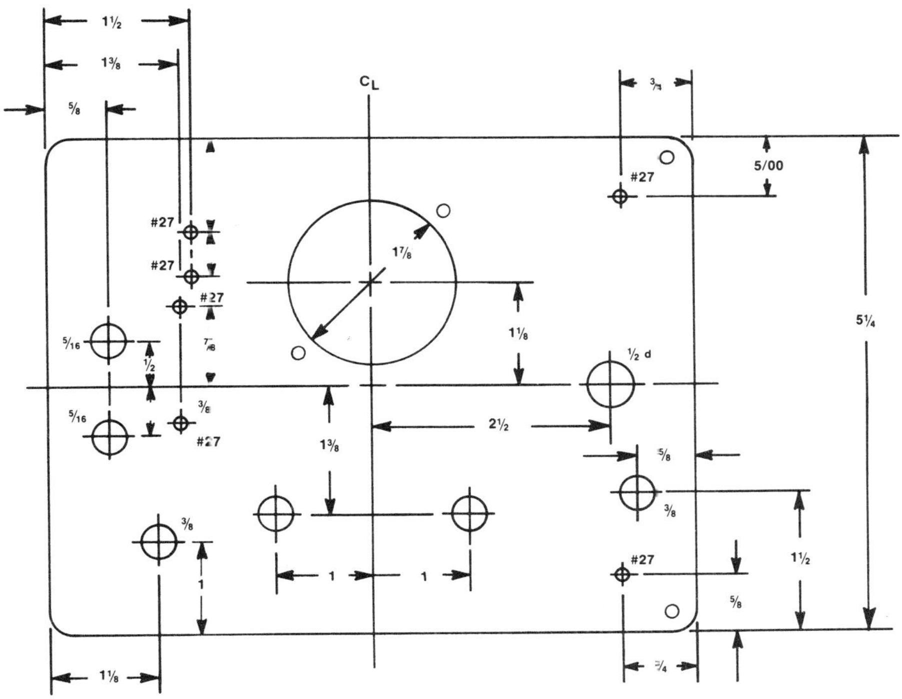

Figure 10-21—Harmonic Analyzer Panel Drilling Layout

PROJECT: Signal Generator for Harmonic Analyzer

While any low distortion, tunable audio signal generator may be used as the input for distortion measurements with the harmonic analyzer described above, it is convenient to have a signal generator designed and built specifically for such use.

The basis for the generator is the very simple Wein Bridge oscillator. The frequency of the Wein Bridge oscillator is determined by two capacitors and two resistors. The gain is provided by a 741 op amp. In order for the oscillator to provide a low distortion sine wave output, some means other than amplifier overload must be provided for limiting the amplitude of oscillation to the linear part of the amplifier characteristic. This limiting and the attendant low distortion is accomplished by means of an automatic gain control. In order to make sure that the system is working properly, an output meter protected by back-to-back LED's is provided across the generator output.

The output frequencies of the generator must be accurately set to match the selection characteristics of the harmonic analyzer. Since using fixed components would be impractical, fine adjustment of each output frequency is provided. The two oscillator capacitors are fixed, while the resistors are all adjustable. The adjustable resistors are printed circuit-board type potentiometers mounted opposite openings in the front panel so that the frequencies can be set and touched up at any time without opening the case. The pairs of resistors are switched by means of a two-pole rotary switch. Another adjustment is necessary in a low distortion Wein Bridge oscillator and that is the feedback circuit of the op amp. This is necessary to keep oscillation amplitude within the area that the AGC can handle effectively. The adjustable feedback resistor is also mounted on the front panel.

PROJECT: Construction of Signal Generator

Figure 10-22 is a complete schematic circuit diagram of the Signal Generator. The variable resistors are mounted on a $1\frac{1}{2} \times 4\frac{1}{2}$ perf board (0.1×0.1) as shown in Figure 10-23 (rear view). Holes in the front panel are provided opposite the adjustment slots of these resistors. Substantially all the other components are mounted on the lower end of the board as shown in Figure 10-24 (rear view). Exact placement of these parts is left up to the experimenter, since

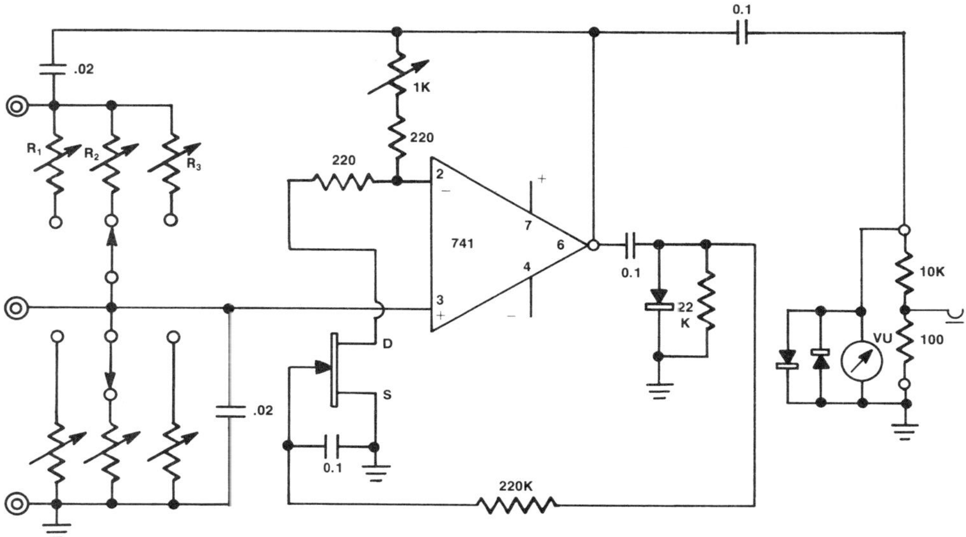

Figure 10-22—Schematic of the Signal Generator

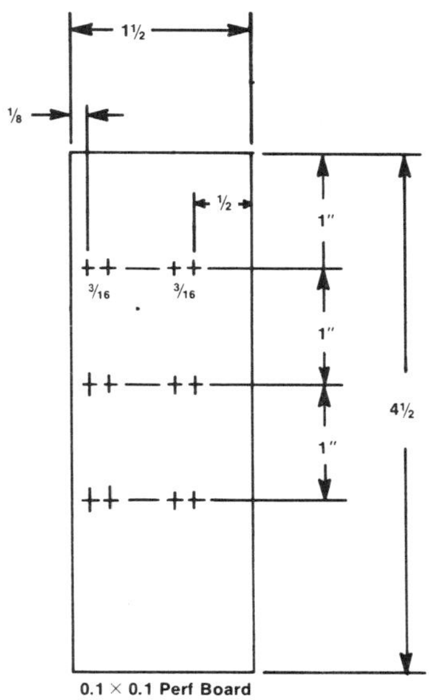

Figure 10-23—Rear View of Positioning of Potentiometer Terminals

part of the skill of an experimenter is the placement of components, and logical wiring of colored leads is shown and provides a convenient means for interconnecting with other components.

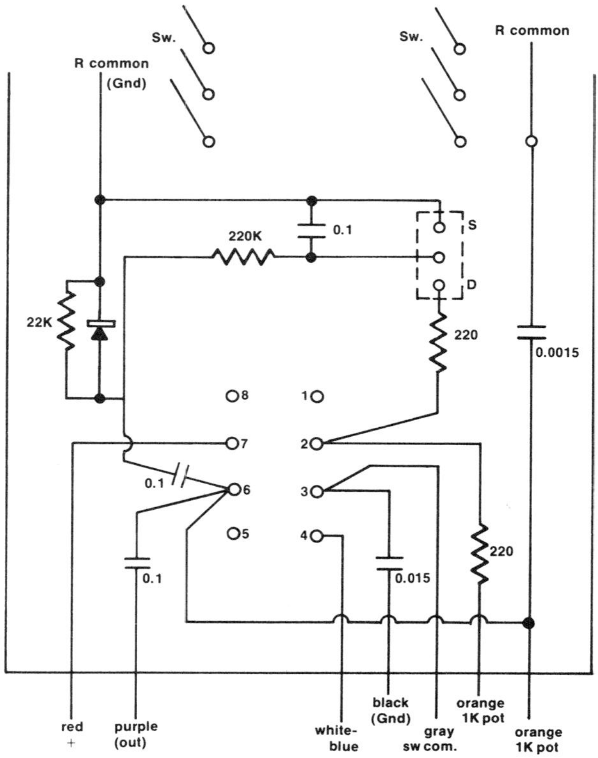

Figure 10-24—Enlarged View of Perf Board Showing Component Positioning

Figure 10-25 is a dimensioned drawing of the front panel. The perf board is mounted on ⅞ inch standoff screws. Figure 10-26 is a table of nominal resistance values for the frequency-determining resistors together with the potentiometer values used to provide these resistances. Figure 10-27 is a photo of the completed analyzer and Figure 10-28 is a photo of the back of the front panel.

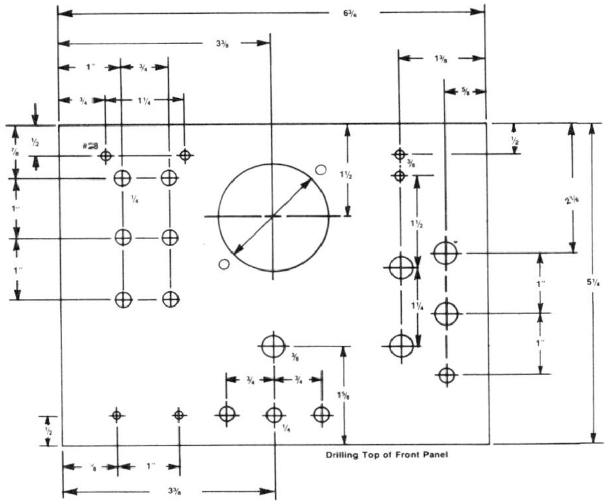

Figure 10-25—Panel Drilling for Top of Panel

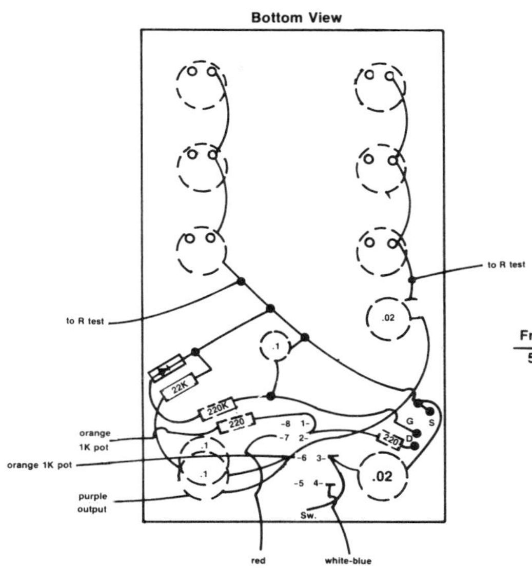

Frequency	R C=0.02μf
5000 Hz	1592 Ω
2500	3184
1667	4777

Figure 10-26—Table of Resistance Values/Potentiometer Values and Layout of Signal Generator

After assembly and wiring, the resistors should be adjusted to the values shown in the table. This is facilitated by the pin-tip jacks on the front panel. Power is supplied through the phone jack on the

Figure 10-27—Harmonic Signal Generator Completed

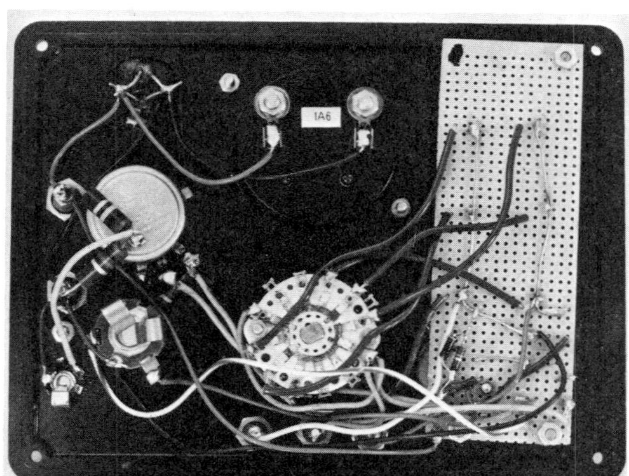

Figure 10-28—Harmonic Signal Generator Inside

front panel and may be from a source of plus and minus 9 volts or, more conveniently, may be derived from the Harmonic Analyzer power supply by means of a two-wire cable terminating in two phone plugs. While no on/off switch has been shown, such may, of course, be added if desired.

The first consideration is that the Generator is operating substantially as expected. If it does not, trouble shooting should be carried out until it does. An Experimenter should not expect every PROJECT to be completed without error. Finding and correcting miswiring, poor soldering, and so forth is part of an Experimenter's continuing experience.

The frequencies of the Generator must be adjusted to precisely match the resonance of the Analyzer. The vernier adjustment on the Analyzer is useful in making frequency adjustments since it can be used to inidcate whether frequencies are high or low.

The procedure is to connect the Generator to the Analyzer and by tuning the Analyzer find whether the Generator frequency is high or low. If it is high, the Generator resistors must be increased and vice versa. Make the resistance adjustments very carefully changing one and then the other of a pair by very small increments. Recheck with an ohmmeter to see that both resistors are close to the same value. When finally adjusted, the 5000, 2500 and 1667 cycle frequencies will accurately represent fundamental, second harmonic and third harmonic respectively. When the frequency determining resistors are correctly adjusted they will be substantially equal, provide the desired frequencies and provide substantially the same amplitude on all frequencies. The feedback potentiometer should be set to provide a 0 DB reading on the output meter. If one or both LED's glow, it indicates either an excessive output voltage or a distorted wave form.

USING THE GENERATOR AND HARMONIC ANALYZER

Two outputs are provided by the Generator, one across the meter which reads 0.8 volt at 0 DB, and the other equals one-thousandth of that or 8 millivolts. If desired, an external attenuator may be used to obtain other values. It is generally assumed that the amplifier or other equipment under test has an input level control.

Connect the Generator to the input of the equipment to be tested and the analyzer to its output. Set the Generator frequency to 5000 cycles. Set the analyzer attenuator to 0. Adjust equipment or analyzer gain to provide a 0 DB reading on the analyzer. Switch to 2500 cycles and advance the analyzer attenuator until a viable reading is obtained. The second harmonic is below the fundamental by the sum of the meter reading and the attenuator setting. For example, if the meter reads 3 at attenuator −20, the second harmonic is -23 DB below the fundamental. Switch to 1667 cycles and repeat the process to obtain the value of the third harmonic distortion.

If the equipment is a tape recorder, record a minute or so at 5000, 2500 and 1667 cycles each. Rewind and play back to the analyzer. Set the gain at 5000 cycles to read zero on the meter. Check the vernier frequency control to provide maximum reading in case the playback speed is slightly different from the recording speed. When the recorder reaches the 2500 cycle section, turn the analyzer attenuator until a viable meter reading is obtained and read the second harmonic distortion. Repeat for the 1667 cycle, third harmonic section.

Various input levels are best examined by using a calibrated attenuator between the generator and the equipment under test.

Glossary

Ampere *The current which flows when one volt is impressed across one ohm. Direct current symbol I; alternating current symbol i.*

Anode *The positive element of a diode or solid state rectifier.*

Base *The control element of a bipolar transistor, which is a transistor employing both positive and negative charge carriers. The most general type of transistor where the base (control) is located between the collector and emitter.*

Cathode *The negative element of a diode or solid state rectifier.*

Collector *The high-voltage element of a bipolar transistor, the current to which is controlled by the base current.*

Decade *A ratio of ten times.*

Diac *A PNP two-terminal device comprising three layers, which breaks down when the applied voltage in either direction exceeds its "breakover voltage." Used as a triggering device as with an SCR or triac.*

Diode *A two-element solid state device which passes current readily in one direction but blocks it in the other. A rectifier.*

Emitter *The low voltage or reference element of a bipolar transistor.*

FET *Field effect transistor.*

Foot-candle *The light intensity at one foot from a standard candle.*

Gate *The control element of a silicon controlled rectifier (SCR) or a field effect transistor (FET).*

Henry *An inductance which exhibits a drop across it of 1 volt when a current of 1 ampere varying at the rate of 1 Hz is passed through it.*

IGFET *Insulated gate field effect transistor.*

JFET *Junction field effect transistor.*

LCD *Liquid crystal display. An applied voltage causes it to change its reflectivity. Requires little exciting current but is visible only by reflecting light. It does not glow like an LED.*

LED *Light emitting diode. Glows brightly when current is passed through it and is available in several colors.*

ma *Milliamperes or thousandths of an ampere.*

Micro- *One millionth of its suffix; -farads, -amperes or -henrys.*

Milli- *One thousandth of its suffix; -amperes or -henrys.*

MOSFET *Metal-oxide-semiconductor field effect transistor.*

mw/cm2 *Milliwats per square centimeter. A measure of radiated energy as received in the form of sunlight.*

N-channel FET *A field effect transistor in which the body or channel is N-type semiconductor material, which is doped to provide excess electrons.*

NPN *A bipolar transistor comprising a layer of P-type material, its base, sandwiched between an N-type material collector and an N-type material emitter.*

N-type material *Semiconductor material such as silicon doped with an impurity to produce an excess of electrons.*

Opto-coupler *An optical coupling device having a light producing input such as a light emitting diode and a light sensitive output such as a light sensitive diode* *(photodiode). Used for coupling signals without a conductive path such as a wire. Used to bridge large potential differences, break a common ground and so forth.*

p *Prefix meaning a millionth of a millionth, standing for pico. For example: pa picoampere, pf picofarad and so forth.*

P-channel FET *A field effect transistor in which the body or channel is P-type semiconductor material, doped to provide an excess of holes, the positive charge counterpart of electrons which are negative.*

Photodiode *A light sensitive diode; a diode the reverse-bias conduction of which increases with increasing incident light.*

Phototransistor *A light sensitive transistor (open base) which becomes more conductive with increasing incident light. More sensitive than the photodiode due to the transistor current gain, however, slower in response.*

PIV *Peak inverse voltage, a rating applied to rectifiers.*

PN *A diode or rectifier in which P-type material is interfaced with N-type material.*

PNP *A bipolar transistor comprising a layer of N-type material sandwiched between a P-type material collector and a P-type material emitter.*

P-type material *Semiconductor material such as pure silicon, doped with an impurity to produce an excess of holes or positive charge carriers.*

RFI *Radio frequency interference; noise or static.*

r.p.m. *Revolutions per minute.*

r.p.s. *Revolutions per second.*

SCR *Silicon controlled rectifier. A four-layer semiconductor device which suddenly conducts between anode and cathode when more than a critical voltage and current are applied between its gate and cathode and which tends to continue to conduct until the anode to cathode voltage is reduced to zero.*

Triac *A five layer semiconductor device similar to an SCR but which can be triggered to conduct in either direction, often by means of a diac as the triggering device.*

μa *Microampere or one millionth of an ampere.*

Zener diode *A PN diode which conducts in the reverse-bias direction whenever more than a critical voltage, the zener voltage, is applied.*

INDEX